Advanced Placement Calculus

Textbook

J. E. Koenka   R. F. Allen

Note for Librarians: A cataloguing record for this book is available from Library and Archives
Canada at www.collectionscanada.ca/amicus/index-e.html
ISBN 1-4120-5874-0

*Printed in Victoria, BC, Canada. Printed on paper with minimum 30% recycled fibre. Trafford's print shop
runs on "green energy" from solar, wind and other environmentally-friendly power sources.*

TRAFFORD

Offices in Canada, USA, Ireland and UK

This book was published *on-demand* in cooperation with Trafford Publishing. On-demand
publishing is a unique process and service of making a book available for retail sale to the
public taking advantage of on-demand manufacturing and Internet marketing. On-demand
publishing includes promotions, retail sales, manufacturing, order fulfilment, accounting and
collecting royalties on behalf of the author.

Book sales for North America and international:
Trafford Publishing, 6E–2333 Government St.,
Victoria, BC v8t 4p4 CANADA
phone 250 383 6864 (toll-free 1 888 232 4444)
fax 250 383 6804; email to orders@trafford.com
Book sales in Europe:
Trafford Publishing (uk) Ltd., Enterprise House, Wistaston Road Business Centre,
Wistaston Road, Crewe, Cheshire cw2 7rp UNITED KINGDOM
phone 01270 251 396 (local rate 0845 230 9601)
facsimile 01270 254 983; orders.uk@trafford.com
Order online at:
trafford.com/05-0775

10 9 8 7 6

TABLE OF CONTENTS

Although this text is intended for Advanced Placement students, it has been written with the purpose of simplifying Calculus, presenting ideas in an as intuitive way as possible and avoiding complex notation. This text is also eminently suitable for International Baccalaureate (Higher Level), 'A' levels and first year Calculus courses.

The mystique of Calculus is such that many students in high school are dissuaded from studying it. The text attempts to convince those students that Calculus is not as mystifying as they believe. For example, limits are not presented until Chapter 4 using a graphical introduction to their study, much use is recommended for the graphing calculator and many questions are designed to generate a hands-on approach to solutions.

The authors have a wealth of experience teaching Calculus and many of the questions have never been seen in textbooks before, derived and designed from classroom work over many years. A few questions have been taken from Advanced Placement exams.

The authors wish to thank Sara Haider, a former student, for her tireless efforts to transpose these ideas into coherent text, diagrams, and graphs.

The authors would be very pleased for teachers who buy books to contact them at jack_koenka@havergal.on.ca or allen@interware.net. We will gladly give free advice or assistance.

J.E. Koenka

R.F Allen

April 2006

CHAPTER 1

Introduction to Slope of a Curved Line

Consider the function $f(x) = 3x$. At each point on the graph, the slope is 3. In other words, the slope is a constant.

A fundamental difference between $f(x) = 3x$ and (say) $g(x) = x^2$ is the fact that the slope of $g(x) = x^2$ varies depending upon the position on the graph.

If one considers a person "walking" from point A to point F on the graph $g(x) = x^2$ as shown:

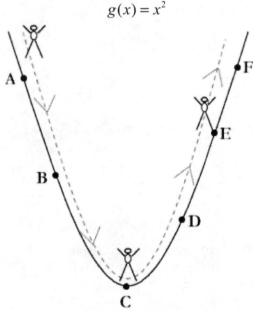

$$g(x) = x^2$$

it is clear that the slope at A is negative, whereas at point E the slope is positive. Furthermore, the more the person walks to the right, the steeper the graph gets (i.e. the slope increases from point C to point F).

In fact, as will be proven and explained later, it turns out that the slope of

$g(x) = x^2$ at a particular point is numerically <u>twice</u> the x co-ordinate of the point.

This means for example that at the point (3,9) the slope is 6.

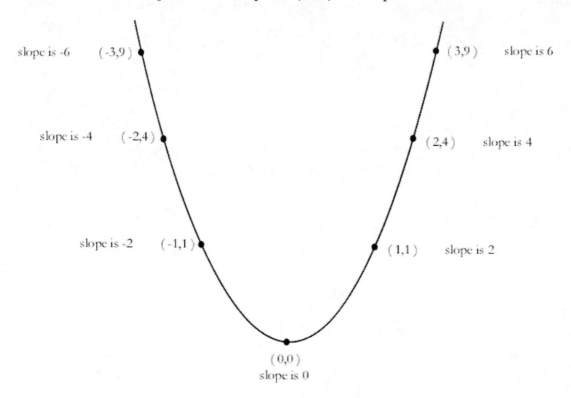

Of course, we need to define what we mean by the slope of a curved line at a point.

<u>Definition</u>

The slope of a smooth curved line at a point is the slope of the tangent at that point,

i.e. in the figure shown:

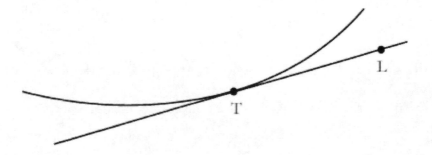

The slope of the curved line at point T is the slope of the straight line TL. Note that we talk about the slope of a curve at a point rather than the slope of a point (which makes little sense).

Example 1

Question: Find an equation of the tangent to $y = x^2$ at the point (4,16).

Answer: The slope of $y = x^2$ at (4,16) is 8 (twice the x co-ordinate).

Using the point–slope form for an equation of a straight line:

$$y - y_1 = m(x - x_1)$$

We get: $y - 16 = 8(x - 4)$

i.e: $y = 8x - 16$

Example 2

Question: $y = 6x + k$ is an equation for a tangent to the graph $y = x^2$. Find the point of tangency P and the value of k.

Answer: $y = 6x + k$ has a slope of 6. Since the slope of $y = x^2$ at a point is twice the x co-ordinate of the point, then the point of tangency must have an x co-ordinate of 3. Therefore, the point of tangency P is (3,9).

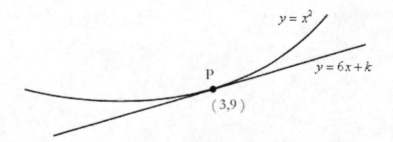

Since P also lies on the tangent $y = 6x + k$ then $9 = 6(3) + k$
i.e. $k = -9$

Notation

The notation commonly used to indicate slope depends upon the form in which the original function is written.

If we write $y = x^2$ then the slope is designated by $\dfrac{dy}{dx}$.

i.e. if $y = x^2$ then $\dfrac{dy}{dx} = 2x$ (twice the x co-ordinate of the point).

If the equation if written in the form $f(x) = x^2$ then the slope is written as $f'(x)$.

i.e. if $f(x) = x^2$ then $f'(x) = 2x$.

Note that f' is a function in its own right. It follows that $f'(3)$, for example, refers to the slope of the curve $y = f(x)$ at the point whose x co-ordinate is 3.

Consider $f(x) = x^2 + 3x + 4$.

This can be thought of as the addition of the function x^2 and the function $3x + 4$.

The slope of the former function is $2x$ and the slope of the latter function is 3. We therefore say that the slope of $f(x) = x^2 + 3x + 4$ is $2x + 3$.

We write: $\quad f(x) = x^2 + 3x + 4$

$\qquad \therefore \; f'(x) = 2x + 3$

In fact it turns out to be true that if

$$f(x) = g(x) + h(x) \qquad \text{then}$$

$$f'(x) = g'(x) + h'(x) \ .$$

The result will be self evident in later work.

Example 3

Question: Find an equation of the tangent to $y = x^2 + 5x + 6$ at point P whose

x co-ordinate is 1.

Answer: Since P lies on the curve then the y co-ordinate of P is:

$$1^2 + 5(1) + 6$$

i.e. P is $(1, 12)$

Also, since $\qquad y = x^2 + 5x + 6$

then $\qquad \dfrac{dy}{dx} = 2x + 5$

$\therefore$ Slope of the tangent at P$(1, 12) = 2(1) + 5 = 7$

$\therefore$ An equation of the tangent at P is

$$y - 12 = 7(x - 1)$$

i.e. $\quad y = 7x + 5$

_____

Example 4

Question: $y = 5x + 6$ is a tangent to $y = x^2 + 3x + c$. Find the point of tangency P

and the value of c.

Answer:

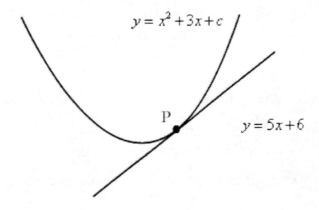

The slope of the tangent at P is 5.

$\therefore$ The slope of $y = x^2 + 3x + c$ at P is also 5.

$\therefore$ $2x + 3 = 5$ where x, remember, denotes the x co-ordinate of P.

i.e. the x co-ordinate of P is 1.

Since P lies on the tangent then the y co-ordinate of P is $5(1) + 6 = 11$.

$\therefore$ P is $(1, 11)$

But P lies on the curve $y = x^2 + 3x + c$ and therefore

$11 = 1^2 + 3(1) + c$

$\therefore$ $c = 7$

Consider $f(x) = 6x^2$.

The slope of the function is 6 times the slope of $y = x^2$. In fact, as again, it will be

clear from later work, if $f(x) = cg(x)$ where c is a constant, then $f'(x) = cg'(x)$.

From this it can be deduced that if $f(x) = 6x^2$ then $f'(x) = 12x$ (i.e. 6 times $2x$).

Note that, for example, $f'(2)$ would be 24.

It follows that if $\quad f(x) = ax^2 + bx + c \quad$ then $\quad f'(x) = 2ax + b$

Another important result which will be proven later is that

if $y = x^n$ then $\dfrac{dy}{dx} = nx^{n-1}$.

For example, if $\quad y = 7x^5 + 2x^3 - 5x + 3$

then $\quad \dfrac{dy}{dx} = 35x^4 + 6x^2 - 5$

Example 5

Question: Find an equation of the tangent to $f(x) = x^3 + 2x^2 - 5x + 1$ at the point

T whose x co-ordinate is 2. Find where this tangent meets the

curve again.

Answer: $f(x) = x^3 + 2x^2 - 5x + 1$ and the y co-ordinate of T is $f(2)$ which

equals $(2)^3 + 2(2)^2 - 5(2) + 1$, which is 7. $\quad \therefore$ T is $(2, 7)$

Also, $f'(x) = 3x^2 + 4x - 5$, therefore at point of tangency T,

$f'(2) = 3(2)^2 + 4(2) - 5 = 15$.

$\therefore$ an equation of the tangent at T is $y - 7 = 15(x - 2)$

i.e. $y = 15x - 23$

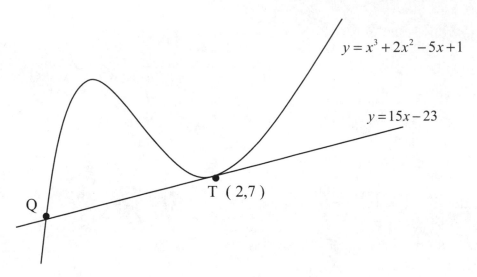

At point Q, where the tangent intersects the curve again:

$$x^3 + 2x^2 - 5x + 1 = 15x - 23$$

i.e. $x^3 + 2x^2 - 20x + 24 = 0$

We know that point T satisfies the equation because T is also a

point where the tangent intersects the curve,

$\therefore$ $x = 2$ is a root of the above equation $x^3 + 2x^2 - 20x + 24 = 0$.

In fact, $x = 2$ is a <u>double root</u> since, in effect, the tangent meets the

curve <u>twice</u> at point T.

Solving $x^3 + 2x^2 - 20x + 24 = 0$

yields $(x - 2)(x - 2)(x + 6) = 0$

$\therefore$ Q is (-6,-113)

Example 6

Question: Find equation(s) of the tangent(s) to $y = x^2 - 2x$ which pass through the point R(4,7).

Answer:

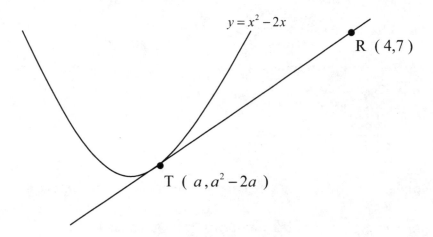

Let the point of tangency T be parametrised as ($a, a^2 - 2a$).

Since $y = x^2 - 2x$ then $\dfrac{dy}{dx} = 2x - 2$.

At point T, the slope is therefore $2a - 2$.

Also, the slope of tangent RT is $\dfrac{rise}{run} = \dfrac{7 - (a^2 - 2a)}{4 - a}$.

$\therefore \qquad 2a - 2 = \dfrac{7 - (a^2 - 2a)}{4 - a}$

i.e. $\qquad -2a^2 - 2 + 10a - 8 = 7 - a^2 + 2a$

i.e. $\qquad 0 = a^2 - 8a + 15$

$\qquad\qquad 0 = (a - 3)(a - 5)$

$\qquad \therefore \ a = 3 \text{ or } 5$

$\therefore$ T is either (3,3) or (5,15).

When T is (3,3) slope = 4, and hence equation of tangent is

$$y - 3 = 4(x - 3)$$

i.e. $\quad y = 4x - 9$

When T is (5,15) slope = 8, and hence equation of tangent is

$$y - 15 = 8(x - 5)$$

i.e. $\quad y = 8x - 25$

Students familiar with the use of graphing calculators are encouraged to use them to check the answers of some of the questions in Worksheet 1.

Note that questions marked with an asterisk * can be checked and/or solved using graphing calculators.

Worksheet 1

SLOPES OF TANGENTS, DERIVATIVES OF POLYNOMIALS

1. Find the equation of the tangent to $y = x^2$ at (3,9). Does the tangent intersect the curve again?

2. Find the equation of the tangent to $y = x^2$ that has a slope of 4.

3. Find the value(s) of a so that the tangents to $y = x^2$ at (a, a^2) and ($-a, a^2$) intersect at 90°.

*4. Find the equation of the tangent to $y = x^3$ at the point (-1,-1). Does the tangent intersect the curve again; if so, where?

5. $y = k$ is a tangent to $y = x^2 + 2x$. Find the value(s) of k.

6. Find the equation of the tangent to $y = x^2 - 3x$ which intersects the x-axis at 45°.

*7. $y = 2x - 1$ is a tangent to $y = x^2$. Name the point of tangency.

8. Find the equations of the tangents to $y = x^2$ which pass through the point (2,3).

9. If $f(x) = x^3$ find $f'(2)$.

*10. Find the equation of the tangent to $y = x^4 - x^2$ at the point where $x = 1$. Does the tangent intersect the curve again?

*11. Find the lowest point on the graph of $y = x^4 - 4x + 1$.

12. Find the equation(s) of the tangent(s) to $y = x^2 + x$ which pass through $(1,1)$.

Answers to Worksheet 1

1. $y = 6x - 9$ No

2. $y = 4x - 4$

3. $a = \dfrac{1}{2}$ or $-\dfrac{1}{2}$

4. $y = 3x + 2$.   Yes, at $(2,8)$.

5. $k = -1$

6. $y = x - 4$

7. $(1,1)$

8. $y = 2x - 1$ or $y = 6x - 9$

9. 12

10. $y = 2x - 2$.   No.

11. $(1,-2)$

12. $y = x$ or $y = 5x - 4$

The process of finding the slope of a curve is called DIFFERENTIATION and the result obtained is called the DERIVATIVE. For example, this means that $2x$ is the derivative of x^2 and we say "differentiating x^2 gives us $2x$".

This is written $D[x^2] = 2x$.

In later work it will be necessary to differentiate more than once.

For example, if $\qquad f(x) = x^3 + 2x^2 + 3x + 5$

$\qquad$ then $\quad f'(x) = 3x^2 + 4x + 3$

$\qquad$ and $\quad f''(x) = 6x + 4$.

$f''(x)$ is called the second derivative of $f(x)$. Its graphical significance will be shown later.

Using the y notation we write:

$$y = x^3 + 2x^2 + 3x + 5$$

$$\therefore \qquad \frac{dy}{dx} = 3x^2 + 4x + 3$$

$$\text{and} \qquad \frac{d^2y}{dx^2} = 6x + 4$$

Note that the location of the "2"s is intentional. It is not a typographical error. The second derivative $\dfrac{d^2y}{dx^2}$ is written in this way to distinguish it from $\left(\dfrac{dy}{dx}\right)^2$. In other words the second derivative (with respect to x) is written $\dfrac{d^2y}{dx^2}$.

Worksheet 2

DERIVATIVE OF x^n

1. Differentiate the following:

   a) $x^5 + 10x^4 - 5x^2 + 3$

   b) $3x^{\frac{1}{2}} - x^{\frac{1}{4}} + 2\sqrt{x} - 3$

   c) $(1-2x)^2$

   d) $\dfrac{1}{2x^2}$

   e) $2x\sqrt{x} - x$

2. Find $\dfrac{dy}{dx}$ for each of the following

   a) $y = (x-3)^2$

   b) $xy = 2$

   c) $2 + x^2 = 3y$

3. Find $f'(x)$ if

   a) $f(x) = 2x^3 + 3x^2 - \dfrac{1}{x}$

   b) $f(x) = \left(1 - \dfrac{1}{x}\right)^2$

*4. Find the points on the graph of $y = x + \dfrac{4}{x}$ where the tangent is horizontal.

5. Find equation(s) of the tangent(s) to the graph of $y = x + \dfrac{1}{x}$ which are parallel to $4y = 3x - 1$.

6. If $f(x) = \dfrac{2 + x^2}{x}$, find $f'(2)$.

7. Find the slope of the tangent to $y = 3x^2 + 4x + 1$ at ($2, 19$).

8. Find the equation of the tangent to $y = 3x^2 - 4x + 2$ at the point where $x = 2$.

*9. $y = x^2 + x$ has a tangent whose equation is $y = 5x - 4$. Find the point of tangency.

10. $y = 5x^2 + 3x + k$ has a tangent whose equation is $y = 13x + 8$. Find the value(s) of k.

11. Find the equation(s) of the tangents to $y = x^2$ which pass through the point ($3, 5$).

12. Find the value(s) of c and a so that $y = cx^3 - 2x^2 + 3x$ has a tangent at the point ($1, a + 1$) whose slope is 5. Find where the tangent meets the curve again.

13. Find the slope of the tangent to $f(x) = (1 + 2x)^2$ at the point ($1, 9$).

14. $y = k$ is a tangent to $y = x^2 + 4x + 2k$. Find the value(s) of k.

15. Find the equation of the tangent to $xy = 3$ at ($3, 1$).

*16. Find the point(s) on the graph of $y = x^3 + x^2 - 5x + 6$ where the tangent is horizontal. Using these points, determine which points are relative maximum and which are relative minimum. (Note that a relative maximum means a point which is higher than all other points in that neighbourhood.)

17. Differentiate $(2x+1)^3$.

18. Show in detail why $y = 3x - 2$ is a tangent to $y = x^3$.

19. If $f(x) = \dfrac{3x+1}{x^2}$, find $f'(3)$.

20. A tangent to $y = x^3$ at the point P meets the curve again at point Q (-2,-8).

Find the co-ordinates of point P.

Answers to Worksheet 2

1. a) $5x^4 + 40x^3 - 10x$

    b) $\dfrac{3}{2}x^{-\frac{1}{2}} - \dfrac{1}{4}x^{-\frac{3}{4}} + x^{-\frac{1}{2}}$

    c) $-4 + 8x$

    d) $-1x^{-3}$

    e) $3x^{\frac{1}{2}} - 1$

2. a) $2x - 6$

    b) $-2x^{-2}$

    c) $\dfrac{2x}{3}$

3. a) $6x^2 + 6x + \dfrac{1}{x^2}$

    b) $2x^{-2} - 2x^{-3}$

4. $(2,4)$ and $(-2,-4)$

5. $4y = 3x + 4$ or $4y = 3x - 4$

6. $\dfrac{1}{2}$

7. 16

8. $y = 8x - 10$

9. $(2,6)$

10. $k = 13$

11. $y = 2x - 1$ and $y = 10x - 25$

12. $c = 2$, $a = 2$. $(-1,-7)$

13. 12

14. $k = 4$

15. $3y + x = 6$

16. $(1,3)$ is a relative minimum

    $(-\dfrac{5}{3}, \dfrac{337}{27})$ is a relative maximum

17. $24x^2 + 24x + 6$

19. $-\dfrac{11}{27}$

20. $(1,1)$

Increasing Functions

A function $f(x)$ is said to be increasing if it is "going uphill".

This means that if the slope of the tangent to the curve at a given point is positive,

then the function is increasing at that point.

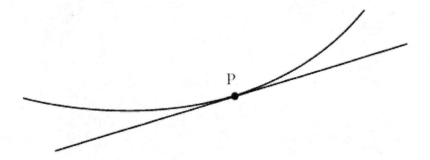

f is increasing at point P because the slope of the tangent at P is positive.

In algebraic terms this means that f is increasing at a point ($a, f(a)$)

if $f(a^+) > f(a)$ where a^+ means values "just a little greater than a".

In summary, f is said to be increasing at a point ($a, f(a)$) if $f'(a)$ is positive.

Similarly, f is decreasing at ($a, f(a)$) if $f'(a)$ is negative.

A function is increasing if its derivative is positive.

This is an important result that we will use later.

Example 7

Question: Find the minimum value of y on the graph of $y = x^2 - 8x + 21$.

Answer:

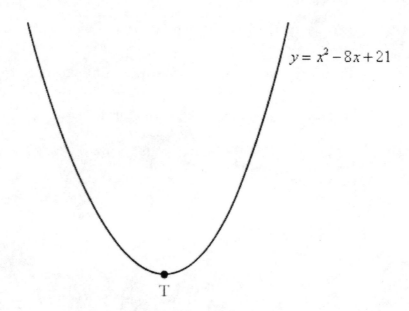

$$y = x^2 - 8x + 21$$

From the graph it is clear that the minimum value of y occurs at point T

where the slope is zero.

$$\frac{dy}{dx} = 2x - 8$$

$\therefore$ At point T, $x = 4$ and T is $(4,5)$.

$\therefore$ The minimum value of y is 5.

Example 8

Question: Find the values of x for which $f(x) = x^3 - 6x^2 + 9x$ is a decreasing

function.

Answer: It is useful to look at the graph of $f(x) = x^3 - 6x^2 + 9x$.

Factoring, we get $f(x) = x(x-3)^2$, from which, by graphing calculator

or previous knowledge we know this looks like:

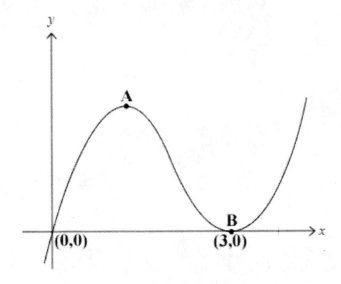

Clearly the function is decreasing from A to B. It is important to

understand that, at points A and B, the slope is zero.

$$f'(x) \quad = 3x^2 - 12x + 9$$

$$= 3(x-1)(x-3)$$

When $x = 1$ or $x = 3$, the slope, $f'(x)$, is zero.

Clearly A has an x co-ordinate of 1 and

     B has an x co-ordinate of 3.

$\therefore \quad f$ is decreasing when $1 < x < 3$.

Normal to a Curve

A line perpendicular to a tangent at the point of tangency is called a normal to a

curve.

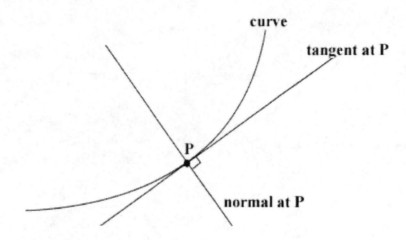

Example 8

Question: Find an equation of the normal to $f(x) = x^3 + \dfrac{1}{x}$ at the point where

$x = 1$.

Answer: $f(x) = x^3 + \dfrac{1}{x} = x^3 + x^{-1}$

Differentiating: $\therefore$ $f'(x) = 3x^2 - 1x^{-2} = 3x^2 - \dfrac{1}{x^2}$

$\therefore$ $f'(1) = 2$.

At point of tangency slope of tangent is 2.

$\therefore$ The slope of the normal is $-\dfrac{1}{2}$.

Also, when $x = 1$, $y = 2$.

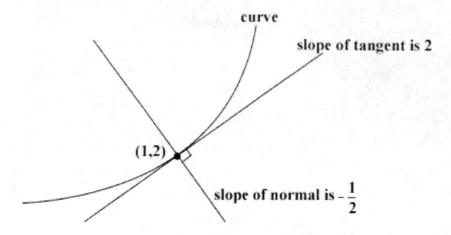

$\therefore$ an equation of the normal is $y - 2 = -\dfrac{1}{2}(x-1)$.   i.e. $x + 2y = 5$

Worksheet 3

INCREASING FUNCTIONS

1. What does it mean to say about the slope of a graph if the function is increasing? For what values of x is $f(x) = 2x^3 - 9x^2 + 12x$ an increasing function?

*2. Find the maximum value of y on the graph of $y = 5 + 3x - x^2$.

3. Find the equation of the normal to $y = \dfrac{1}{x}$ at $(2, \dfrac{1}{2})$.

4. Find the equation of the normal to $y = (x-1)(x-3)$ at $(3, 0)$.

5. For what values of x is the slope of $f(x) = x^3 - 3x^2$ positive?

6. For what values of x is the slope of $f(x) = \dfrac{x^3}{3} - \dfrac{3x^2}{2} + 2x + 1$ positive?

*7. Let $f(x) = \dfrac{x^4}{2} + 2x^3 + 2x^2 + 8$. For what values of x is $f(x)$ increasing? For what values of x is $f(x)$ decreasing? Use these results to decide which points are relative maxima and which are relative minima.

8. Show in detail that $y = \dfrac{1}{x}$ and $y = 3x - 2x^2$ are tangent to one another.

9. Find the point(s) on the graph of $y = 3x^2 - 4x$ where the tangent is parallel to $y = 8x$.

10. Find the value(s) of k if $y = x + k$ is a tangent to $y = x^2 + 3x + 3$.

11. Let $f(x) = \dfrac{x+1}{x^2}$. Explain why the graph has no y intercept. Find where the slope equals zero. Is this point a relative maximum or a relative minimum?

12. Find the value of k if $y = 2x + k$ is a tangent to $y = x^2 + 3x + 3$.

Answers to Worksheet 3

1. $x < 1$ or $x > 2$
2. $\dfrac{29}{4}$
3. $y = 4x - 7\dfrac{1}{2}$

4. $2y + x = 3$
5. $x < 0$ or $x > 2$
6. $x < 1$ or $x > 2$

7. $-2 < x < -1$ or $x > 0$ (f is increasing)

$x < -2$ or $-1 < x < 0$ (f is decreasing)

$(-2, 8)$, $(0, 8)$ are relative minima.

$(-1, 8\dfrac{1}{2})$ is a relative maximum.

9. $(2, 4)$
10. $k = 2$
11. $(-2, -\dfrac{1}{4})$ minimum
12. $k = \dfrac{11}{4}$

Worksheet 4

*1.   Find the points on the graph of $y = 2x^3 - 9x^2 + 12x + 1$ where the tangent is

horizontal. Hence locate a relative maximum point.

*2.   For what values of x is f increasing if $f(x) = 2x^3 - 15x^2 + 36x + 1$.

3.   Find the equation of the normal $y = x^3 + 4x + 1$ at the point where $y = 1$.

4.   Find the equation(s) of the tangent(s) to $y = x + \dfrac{1}{x}$ which are parallel to

$y + 3x = 1$.

5.   The tangent to $y = x^3 + x^2 + x + 1$ at the point $(1,4)$ meets the curve again at

point Q. Find the co-ordinates of Q.

*6.   Find the maximum value of y on the graph of $y = 6 + 4x - x^2$.

7.   Find the value of k if $y = 2x + k$ is a tangent to $y = x^2 + 4x + 1$.

*8.   Find the values of x for which f is decreasing if $f(x) = x^3 - 3x^2 + 3$.

9.   Find the value of c so that $y = x^3 + cx^2 + 3x + 4$ has a tangent at $(-1, c)$

whose slope is 2. Find where the tangent meets the curve again.

10.   Find the points on $y = 2x^3 - 9x^2 - 24x + 1$ where the tangent is horizontal.

11. $y = x^2$ is tangent to $y = -x^2 + kx - k$. Find the value(s) of k and the point(s)

of tangency.

Answers to Worksheet 4

1. $(1,6)$ and $(2,5)$

   $(1,6)$ is a relative maximum point.

2. $x < 2$ or $x > 3$

3. $4y + x = 4$

4. $y + 3x = 4$ or $y + 3x = -4$

5. $(-3,-20)$

6. 10

7. $k = 0$

8. $0 < x < 2$

9. $c = 2$ $(0,4)$

10. $(4,-111)$ and $(-1,14)$

11. $k = 8$ $(2,4)$ or $k = 0$ $(0,0)$

CHAPTER 2

Properties of Derivatives

To investigate derivatives using first principles, we will look at the slope of $f(x) = x^2$

at the point P (3,9).

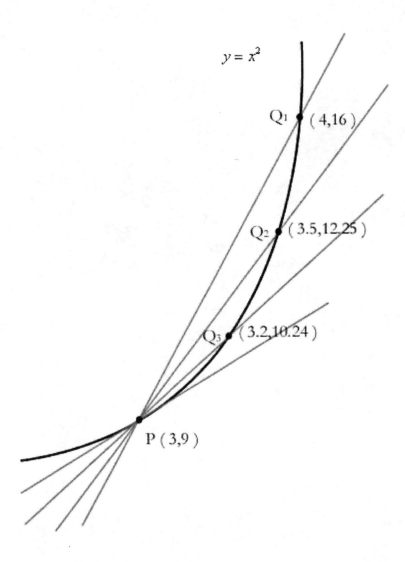

Let Q_1, Q_2, Q_3, Q_4, … be a sequence of points on the curve getting closer and closer

to P. Let Q_1 be (4,16), Q_2 be (3.5,12.25), Q_3 be (3.2,10.24), Q_4 be (3.1,9.61)

The idea is that the tangent to the curve at point P is the limiting line of the sequence of lines PQ_1, PQ_2, PQ_3, PQ_4 ...

We say that the slope of the tangent at P is the limit of the slopes of lines PQ_1, PQ_2, PQ_3...

$$\text{slope } PQ_1 = \frac{16-9}{4-3} = 7$$

$$\text{slope } PQ_2 = \frac{12.25-9}{3.5-3} = 6.5$$

$$\text{slope } PQ_3 = \frac{10.24-9}{3.2-3} = 6.2$$

$$\text{slope } PQ_4 = \frac{9.61-9}{3.1-3} = 6.1$$

It certainly appears that the slope of the tangent at P is "close to" 6.

<u>To show that the slope of $f(x) = x^2$ at (3,9) is 6</u>

Let P be (3,9) and Q be a point $\left(3+h, (3+h)^2\right)$ on the curve $y = x^2$

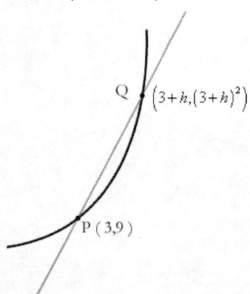

The slope of PQ is $\dfrac{(3+h)^2-9}{(3+h)-3}=\dfrac{6h+h^2}{h}=6+h$. *****

In the "limiting case" i.e. where Q actually coincides with P then the line PQ becomes the tangent at P and h becomes zero.

From the above ***** it can be deduced that when $h=0$, the slope of the tangent i.e. the slope of the limiting case of PQ is 6+0 which is 6.

To find the slope of $f(x) = x^2$ at (a, a^2)

From the previous discussion we can say that the slope of the tangent at P is the limit of the slope of PQ as Q coincides with P i.e. as h approaches zero.

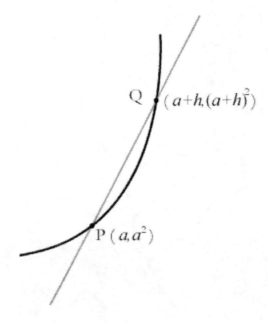

The slope of the tangent at P is written:

$$\lim_{h\to 0}\ \text{slope PQ} \qquad =\lim_{h\to 0}\frac{(a+h)^2-a^2}{(a+h)-a}$$

$$= \lim_{h \to 0} \frac{2ah + h^2}{h}$$

$$= \lim_{h \to 0} 2a + h$$

When, in the limiting case, h <u>does</u> become zero, then the slope of the tangent at

$P = 2a + 0 = 2a$. This means that if $f(x) = x^2$ then $f'(a) = 2a$.

i.e. The slope of $f(x) = x^2$ at an arbitrary point (a, a^2) is $2a$.

$$\boxed{\text{In general } D_x \left(x^2 \right) = 2x}$$

<u>Definition of derivative from first principles</u>

Consider $y = f(x)$

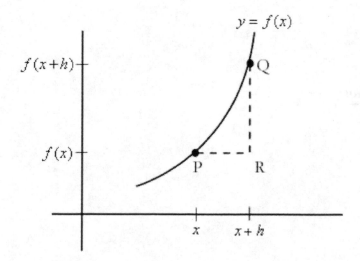

$f'(x)$ is the limit of $\dfrac{QR}{PR}$ as Q coincides with P.

i.e. $\boxed{f'(x) = \lim_{h \to 0} \dfrac{f(x+h) - f(x)}{h}}$

This is called the first principles definition of the derivative of any function, $f(x)$.

If we use the $\dfrac{dy}{dx}$ notation then we say $\dfrac{dy}{dx} = \lim\limits_{\Delta x \to 0} \dfrac{\Delta y}{\Delta x}$.

Example 1

To find the derivative of $\dfrac{1}{x}$ from first principles

$$f(x) = \frac{1}{x}$$

$$f'(x) = \lim_{h \to 0} \frac{f(x+h) - f(x)}{h}$$

$$= \lim_{h \to 0} \frac{\dfrac{1}{x+h} - \dfrac{1}{x}}{h}$$

$$= \lim_{h \to 0} \frac{\dfrac{x - (x+h)}{(x+h)x}}{h}$$

$$= \lim_{h \to 0} \frac{-h}{h(x+h)x}$$

$$= \lim_{h \to 0} \frac{-1}{(x+h)x}$$

$$= \frac{-1}{x^2}$$

Note that from Chapter 1 we learned the rule $D(x^n) = nx^{n-1}$.

Therefore: $D\left(\dfrac{1}{x}\right) = D\left(x^{-1}\right) = (-1)x^{-2} = \dfrac{-1}{x^2}$ as shown.

Example 2

To prove $D(x^n) = nx^{n-1}$ for the cases where n is a positive integer

Let $f(x) = x^n$

Then $f'(x) \triangleq \lim_{h \to 0} \dfrac{f(x+h) - f(x)}{h}$

$$= \lim_{h \to 0} \dfrac{(x+h)^n - x^n}{h}$$

(By binomial theorem) $= \lim_{h \to 0} \dfrac{x^n + \binom{n}{1}x^{n-1}h + \binom{n}{2}x^{n-2}h^2 + \ldots\ldots - x^n}{h}$

$$= \lim_{h \to 0} \binom{n}{1}x^{n-1} + \binom{n}{2}x^{n-2}h + \binom{n}{3}x^{n-3}h^2 + \ldots.$$

$$= \binom{n}{1}x^{n-1}$$

$$= nx^{n-1}$$

CHAIN RULE (to be proven at the end of the chapter)

$$D\big[f(g(x))\big] = f'(g(x)) \cdot g'(x)$$

This result is best illustrated by examples.

The graph of a function is said to be continuous if it is possible to draw the graph without lifting the pencil from the paper. ie if it has no "jumps". It is not possible to find the slope or the derivative of a function at a discontinuity.

Example 3

Consider $(x^2+1)^5$. This function can be decomposed by letting $f(x)=x^5$ and

$g(x)=x^2+1$. Then $f(g(x))=(x^2+1)^5$ as required.

Note that $f'(x)=5x^4$ and $g'(x)=2x$.

$$D\left[(x^2+1)^5\right]=D\left[f(g(x))\right]$$

$$=f'(g(x))\cdot g'(x)$$

$$=5\left[g(x)\right]^4\cdot 2x$$

$$=10x(x^2+1)^4$$

Example 4

$$D\left[(x^2+3x)^{\frac{3}{2}}\right]=\frac{3}{2}(x^2+3x)^{\frac{1}{2}}\cdot(2x+3)$$

$$=\frac{3(2x+3)}{2}\sqrt{x^2+3x}$$

In layman terms, the Chain Rule can be thought of as "when differentiating a complicated expression, differentiate the most outside function, writing down the bracket as it is, then <u>multiply</u> by the derivative of what's inside the bracket."

Example 5

$$D\left[(2x+1)^6\right] = 6(2x+1)^5 \text{ times } 2$$

derivative of 6th power

bracket written
down unchanged

derivative of bracket

Example 6

$$D\left[\sqrt{3x-5}\right] = D(3x-5)^{\frac{1}{2}}$$

$$= \frac{1}{2}(3x-5)^{-\frac{1}{2}} \cdot 3$$

$$= \frac{3}{2\sqrt{3x-5}}$$

Example 7

$$D\left[\frac{1}{(x^2-7)^3}\right] = D(x^2-7)^{-3}$$

$$= -3(x^2-7)^{-4} \cdot 2x$$

$$= \frac{-6x}{(x^2-7)^4}$$

In a longer chain of functions we have

$$D\Big[f\big(g(h(x))\big)\Big] = f'\big(g(h(x))\big)g'(h(x))h'(x).$$

This can be extended to any length chain of functions.

In dy, dx notation the Chain Rule can be expressed as

$$\frac{dy}{dx} = \frac{dy}{dt} \times \frac{dt}{dx}$$

OR $\quad \dfrac{dy}{dx} = \dfrac{dy}{dt} \div \dfrac{dx}{dt}$

In a longer chain we can say

$$\frac{dy}{dx} = \frac{dy}{du} \cdot \frac{du}{dv} \cdot \frac{dv}{dx}$$

i.e. We can think of these as though they were fractions where we can 'cancel'. In fact, they are limits of fractions to be studied in Chapter 4.

We have seen earlier that

$$D\Big[f(x) + g(x)\Big] = f'(x) + g'(x)$$

but it is NOT TRUE that we can extend this result to the derivative of the <u>product</u> of two functions.

<u>PRODUCT RULE</u> (to be proven at the end of the chapter)

Consider $\quad f(x) = g(x)h(x)$

then $\quad f'(x) = h(x)g'(x) + g(x)h'(x)$

Example 8

If $\quad f(x)=\sqrt{x+1}\cdot(x^2+3)$

then $\quad f'(x)=(x^2+3)\cdot\frac{1}{2}(x+1)^{-\frac{1}{2}}+\sqrt{x+1}\cdot2x$

Example 9

$$D\left[(x^2+5)^4\cdot(x^3+6)^7\right]$$

$$=(x^3+6)^7\cdot4(x^2+5)^3\cdot2x+(x^2+5)^4\cdot7(x^3+6)^6\cdot3x^2$$

A useful mnemonic is the following:

If u and v functions of x then

$$D[u\cdot v]=vdu+udv$$

Where du, dv denote the derivatives of u and v respectively with

respect to x.

Similarly, the derivative of a quotient can be obtained by the following formula:

QUOTIENT RULE (to be proven at the end of the chapter)

$$D\left[\frac{f(x)}{g(x)}\right]=\frac{g(x)\cdot f'(x)-f(x)\cdot g'(x)}{\left[g(x)\right]^2}$$

A mnemonic for this is

$$D\left[\frac{u}{v}\right]=\frac{vdu-udv}{v^2}$$

Example 10

$$D\left[\frac{2x+1}{1-3x^2}\right]\quad=\quad\frac{(1-3x^2)2-(2x+1)(-6x)}{(1-3x^2)^2}\quad=\quad\frac{2+6x+6x^2}{(1-3x^2)^2}$$

Worksheet 1

PRODUCT RULE, QUOTIENT RULE, CHAIN RULE, FIRST PRINCIPLES

1. Find $\dfrac{dy}{dx}$ for each of the following:

    a) $y = 16x + 2x^3$

    b) $y = 2x\left(8 + x^2\right)$

    c) $y = (x-1)(x-3)$

    d) $y = \left(x^3 - 1\right)\left(x^2 - 3\right)$

    e) $y = \dfrac{x-1}{x-3}$

    f) $y = \sqrt{2x-1}$

    g) $y = x\sqrt{2x-1}$

    h) $y = (3x-1)^{\frac{1}{4}}$

    i) $y = \sqrt{6x^2 + 2x + 1}$

    j) $y = \sqrt{x^2 + x + 1}$

2. Find $f'(x)$ for each of the following:

    a) $f(x) = 4x^2 - 6x - \dfrac{1}{3x}$

    b) $f(x) = \left(x^2 + 5\right)^5 (1-x)^3$

    c) $f(x) = \dfrac{x^2 - 4x}{2x + 6}$

    d) $f(x) = \sqrt{x^2 - 4}\left(2x - x^2\right)$

3. Find $\dfrac{dy}{dx}$ if $x + xy + y = 2$.

4. $f(x) = \sqrt{x^2 - 1}$. Find $f'(2)$.

5. Find two values of x for which the tangent to $y = \dfrac{2x}{x+1}$ is parallel to

    $y - 2x + 8 = 0$.

6. Given that $D_x\left(x^n\right) = nx^{n-1}$, show by using the Product Rule that

    $D_x\left(x^{2n}\right) = 2nx^{2n-1}$.

7. Find the derivative of $\dfrac{1}{x^2}$ using first principles.

8. If $f'(x)=6x+3$ and $f(1)=7$ find $f(2)$.

9. Find where $y=-3x+x^3$ is decreasing.

10. Prove that $y=\dfrac{x^3}{3}+2x^2+4x$ is always increasing.

11. Find the equation of the tangent to $y=\sqrt{25-x^2}$ at the point $(4,3)$.

Answers to Worksheet 1

1. a) $16+6x^2$

    b) $16+6x^2$

    c) $2x-4$

    d) $5x^4-9x^2-2x$

    e) $\dfrac{-2}{(x-3)^2}$

    f) $\dfrac{1}{\sqrt{2x-1}}$

    g) $\dfrac{3x-1}{\sqrt{2x-1}}$

    h) $\dfrac{3}{4}(3x-1)^{\frac{-3}{4}}$

    i) $\dfrac{6x+1}{\sqrt{6x^2+2x+1}}$

    j) $\dfrac{2x+1}{2\sqrt{x^2+x+1}}$

2. a) $8x-6+\dfrac{1}{3x^2}$

    b) $10x(1-x)^3(x^2+5)^4-3(x^2+5)^5(1-x)^2$

    c) $\dfrac{2x^2+12x-24}{(2x+6)^2}$

    d) $\dfrac{-3x^3+4x^2+8x-8}{\sqrt{x^2-4}}$

3. $\dfrac{-3}{(x+1)^2}$

4. $\dfrac{2}{\sqrt{3}}$

5. $x=0$ OR -2

8. 19

9. $-1<x<1$

11. $3y+4x=25$

38

Worksheet 2

1. Find $\dfrac{dy}{dx}$ for each of the following:

    a) $y = \dfrac{1}{4x^2 + 1}$
        b) $y = (2x+3)^{n+1}$

    c) $y = x^2 y + 1$
        d) $y = x\sqrt{x^2 + 9}$

    e) $y = (x-1)(x-2)(x-3)$

2. Differentiate $(2x^2 + 1)(3x^4 - 3x + 2)$ with respect to x.

3. Find $D_x\left(\dfrac{x^2 + 1}{x + 1}\right)$.

4. Find $\dfrac{dy}{dx}$ for each of the following:

    a) $y = (3x+1)^{\frac{5}{3}}$
        b) $y = 3x\sqrt{x^2 + 1}$

    c) $y = \dfrac{2x - 1}{2x + 1}$
        d) $xy + x^2 y = 3$

5. Find the equation of the tangent to the curve $y = x\sqrt{25 - x^2}$ at $(0,0)$.

    Note that $(0,0)$ is on the graph.

6. Find the equation of the tangent to $x^2 + y^2 = 25$ at $(3,4)$.

7. Differentiate $\sqrt[3]{5 - 6x^2}$.

8. $4y = x + k$ is a tangent to $y = \sqrt{x - 1}$. Find k.

9. Find the maximum value of $\dfrac{2}{3 + 4x^2}$.

10. Find the minimum value of $\sqrt{x^2 + 4x + 5}$.

Answers to Worksheet 2

1. a) $\dfrac{-8x}{\left(4x^2+1\right)^2}$

    b) $2(n+1)(2x+3)^n$

    c) $\dfrac{2x}{\left(1-x^2\right)^2}$

    d) $\dfrac{2x^2+9}{\sqrt{x^2+9}}$

    e) $3x^2-12x+11$

2. $36x^5+12x^3-18x^2+8x-3$

3. $\dfrac{x^2+2x-1}{\left(x+1\right)^2}$

4. a) $5(3x+1)^{\frac{2}{3}}$

    b) $\dfrac{6x^2+3}{\sqrt{x^2+1}}$

    c) $\dfrac{4}{\left(2x+1\right)^2}$

    d) $\dfrac{-3-6x}{\left(x+x^2\right)^2}$

5. $y=5x$

6. $3x+4y=25$

7. $-4x \cdot \left(5-6x^2\right)^{-\frac{2}{3}}$

8. $k=3$

9. $\dfrac{2}{3}$

10. 1

Relative Maximum Minimum Points

A relative maximum point occurs at a point on a curve where the y value at that point is greater than the y values of points in its neighbourhood. Usually this will occur when the slope is zero.

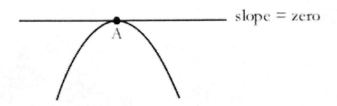

A is a relative maximum point (on a smooth continuous curve).

A relative maximum point occurs when the slope changes from $+$ to $-$.

Example 11

To find a relative maximum point on the graph of $y = x^3 - 6x^2 + 9x + 4$:

$$\frac{dy}{dx} = 3x^2 - 12x + 9$$

$$= 3\left(x^2 - 4x + 3\right)$$

$$= 3(x-1)(x-3)$$

Note that when $x = 1$ or $x = 3$ the slope equals zero.

On a signed line diagram we have:

x		1		3	
$f'(x)$	$+$		$-$		$+$

Note that $f'(x)$ changes from $+$ to $-$ as x "passes through" the value 1.

∴ $x = 1$ yields a relative maximum.

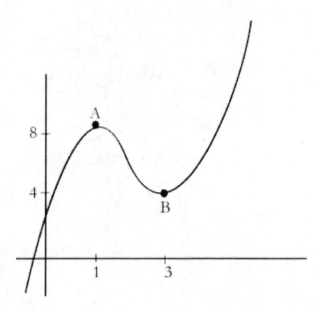

A (1,8) is a relative maximum.

Similarly, note that B (3,4) is a relative minimum point where $f'(x)$ changes from

−   to   +   as x "passes through" 3.

<u>Summary</u>

A relative minimum occurs when $f'(x)$ changes from   −   to   +.

A relative maximum occurs when $f'(x)$ changes from   +   to   −.

An absolute maximum point is a point which is higher than all other points in the

domain of the function, regardless of whether the function is discontinuous or has a

slope of zero at the point. ie (a, f(a)) is an absolute maximum point if f(a)>f(x) for

<u>all</u> values of x in the domain of f.

If a curve is discontinuous however and $f'(x)$ changes sign then there will not be a

relative max/min point. For example, $y = \dfrac{1}{x^2}$ looks like:

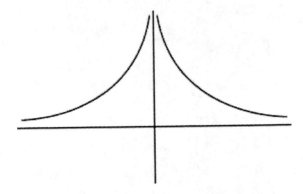

When $x = 0$, $f'(x)$ changes from $+$ to $-$ but there is not a local maximum

point because of the discontinuity.

A critical value of a function is a value such that $f'(x) = 0$ or is undefined changing

from $+\infty$ to $-\infty$ or vice versa.

For example consider $f(x) = \dfrac{x+1}{(x-1)^2}$ which looks like:

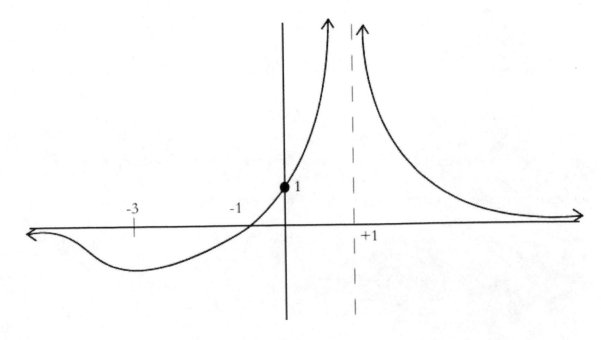

On a sign diagram we have:

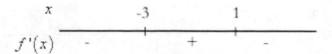

and $f'(x)$ changes sign at both $x = -3$ and $x = 1$ but $x = -3$ yields a relative

minimum whereas $x = 1$ does not yield either a maximum or a minimum because of

the discontinuity. $x = -3$ and $x = 1$ however are critical values.

Note that further notes on relative max/min points are given in Chapter 3.

The Second Derivative

Let's consider the significance of the derivative of f', written f''.

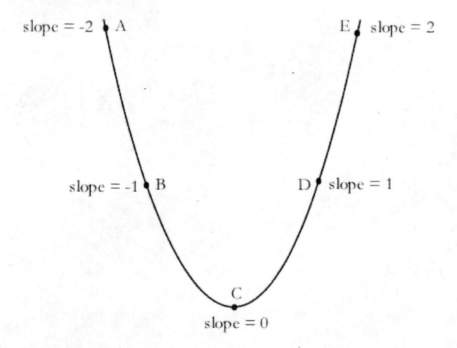

Note that in going from A to E along the curve the slope of the curve is increasing everywhere. The slope of the curve is increasing everywhere i.e. $f'(x)$ is increasing.

From Chapter 1 we know that if a function is increasing then its derivative is positive. It follows that if $f'(x)$ is increasing then $f''(x)$ is positive.

It can therefore be deduced that, for the curve shown, $f''(x)$ is positive everywhere.

 → is called concave up (note the useful mnemonic 'cup')

i.e. $f''(x)$ is positive means the graph is part of a cup.

Similarly, $f''(x)$ is negative implies the graph is concave down.

i.e. like an umbrella:

f'' is negative – umbrella – rain – ☹ – frown – down

f'' is positive – cup – up – ☺ – smile – happy

We say that $f''(x)$ is a measure of the curvature (or concavity) although the numerical relationship is not easy to define. Suffice it to say, usually, the larger the numerical value of $f''(x)$ at a particular point, the more "bendy" the curve.

Consider the graph shown below:

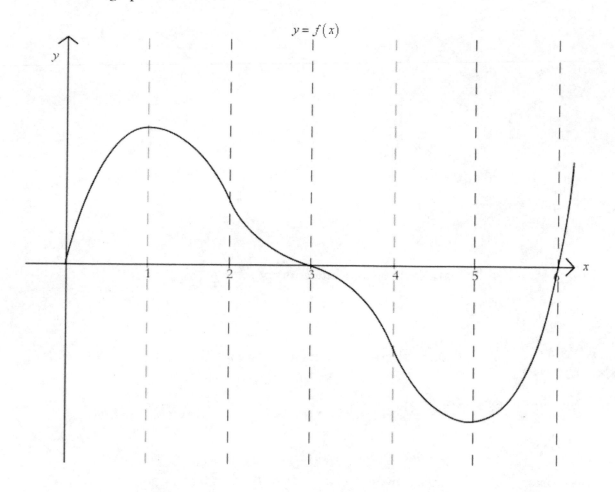

$$y = f(x)$$

Convince yourself that the table below represents the values of f, f' and f'' for the different intervals of x.

	$0 < x < 1$	$1 < x < 2$	$2 < x < 3$	$3 < x < 4$	$4 < x < 5$	$5 < x < 6$
f	$+$	$+$	$+$	$-$	$-$	$-$
f'	$+$	$-$	$-$	$-$	$-$	$+$
f''	$-$	$-$	$+$	$-$	$+$	$+$

Note that in particular when $x = 2$ the graph is changing concavity from down to up

i.e. f'' is changing from negative to positive

i.e. $f''(2) = 0$.

We say that when $x = 2$ the graph has an inflection point.

Inflection Points

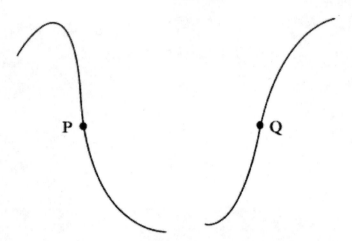

Points P and Q illustrate examples of inflection points.

An inflection point occurs when $f''(x)$ changes sign and when it is possible to draw a tangent at that point.

Most often, but not always, this will occur when $f''(x)=0$. A good method for locating inflection points is to find where $f''(x)=0$ and check to see if f'' changes sign at those values of x.

It is possible for an inflection point to occur when f'' is infinite.

For example, consider $f(x)=x^{\frac{1}{3}}$. At (0,0) the graph looks like → (0,0)

f'' is infinite but (0,0) is an inflection point since $f''(0^-)$ is positive and $f''(0^+)$ is negative.

It is not true however that $f''(x)=0$ guarantees an inflection point. For example, on the graph of $y = x^4$ at $(0,0)$ the second derivative is 0 but $(0,0)$ is not an inflection point because $f''(x)$ does not change sign at $(0,0)$ since $f''(0^-)$ is positive and $f''(0^+)$ is positive also.

Note also that if $f(x)$ is discontinuous when $x = a$ then it is not an inflection point even if $f(x)$ changes concavity because it is not possible to draw a tangent when $x = a$.

For example, $f(x) = \dfrac{1}{x}$ changes concavity when $x = 0$ but $x = 0$ does not yield an inflection point.

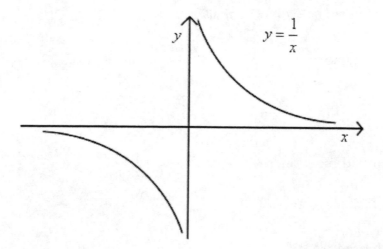

In summary, remember that

An inflection point occurs when $f''(x)$ changes sign
AND it is possible to draw a tangent.

It is quite often helpful to think of an inflection point as a point where the slope is a maximum or a minimum.

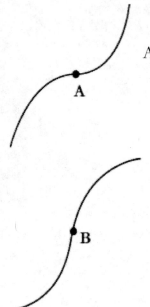

A is an inflection point where slope is a minimum.

B is an inflection point where slope is a maximum.

An even function is one in which

$$f(x) = f(-x) \text{ for all } x$$

An example is

$$f(x) = \frac{2x^4 + 3}{x^2 - 1}$$

An odd function is one in which

$$f(-x) = -f(x)$$

An example is

$$f(x) = x^3 + 4x$$

It follows that an even function is symmetric about the y axis and an odd function is symmetric about the origin.

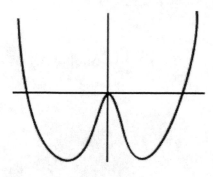

An even function

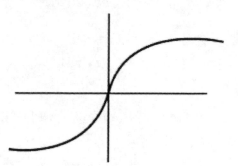

An odd function

Worksheet 3

1. Find the equations of the tangents, with slope 9, to the curve $y = x^2(x-3)$.

2. Show that the function $f(x) = \dfrac{x^3}{3} - 2x^2 + 4x + 3$ is <u>always</u> increasing.

3. The curve $f(x) = \left(a + \dfrac{b}{x}\right)\sqrt{x}$ passes through the point (4,8) at which the slope of the tangent is 2. Find a and b.

4. $y = 2x$ is a tangent to $y = x^2 + 4x + k$. Find k.

5. Prove by first principles that if $f(x) = \sqrt{x}$ then $f'(x) = \dfrac{1}{2\sqrt{x}}$.

6. Find the equation of the curve whose slope at the point (x, y) is $2x + 3$ if the curve passes through (1,6).

7. Find, for what values of x, $f(x)$ is increasing if $f(x) = (x-1)^2(x+2)$.

8. Find where $y = -x^3 + 3x$ is increasing.

9. If $f''(x) > 0$ for all x for a certain function f, what can be said about the graph of f?

10. Find the inflection points of $f(x) = x^4 - 2x^3 + 5x - 3$.

11. Find the minimum points on the graph of $f(x) = x^4 - 8x^2 + 4$.

12. For what values of k will the graph of $y = 3x^3 + 6x^2 + kx + 5$ have positive slope for <u>all</u> values of x?

13. If $f(x) = x^3 - 3x$, find   a) $f(2)$ b) $f'(2)$ c) $f'(-1)$

14. Find the equation of the tangent to $y = \dfrac{1}{x^2} + \dfrac{1}{x^3}$ at the point where $x = 1$.

15. Find the equation of the tangent to $y = 8x^4 + 40x^3 + 26x^2 - 63x + 24$ at the point (-3,15). Find another point on the curve which has exactly the same equation for the tangent at the point.

16. Find a point P on $y = x^3$ such that the tangent at P intersects the original curve again at point Q so that the slope of the tangent at Q is 4 times the slope of the tangent at P.

17. Shown below is a graph of $y = f'(x)$, the derivative of $f(x)$.

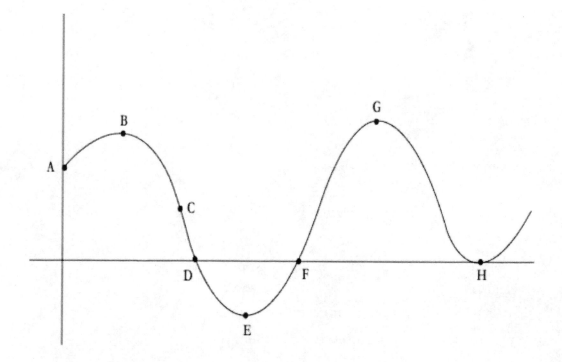

Describe what signifance, if any, points A, B, C, D, E, F, G, H tell us about the graph of $y = f(x)$, the original function.

52

Answers to Worksheet 3

1. $y = 9x + 5$ or $y = 9x - 27$

2. ---

3. $a = 6$, $b = -8$

4. $k = 1$

5. ---

6. $y = x^2 + 3x + 2$

7. $x < -1$ or $x > 1$

8. $-1 \le x \le 1$

9. graph is always concave upwards like a parabola

10. $(0,-3)$ $(1,1)$

11. $(-2,-12)$ $(2,-12)$

12. $k > 4$

13. a)2   b)9   c)0

14. $y = -5x + 7$

15. $y = -3x + 6$ $(\frac{1}{2}, 4\frac{1}{2})$

16. <u>Any</u> point on the curve satisfies the conditions stated.

$\therefore$ P can be any point on $y = x^3$.

17. A tells us $f(x)$ is increasing.

B tells us $f(x)$ has an inflection point like
C tells us $f(x)$ is increasing.
D tells us $f(x)$ has a relative maximum point.

E tells us that $f(x)$ has an inflection point like
F tells us that $f(x)$ has a relative minimum point.

G tells us that $f(x)$ has an inflection point like

H tells us that $f(x)$ has an inflection point like

Worksheet 4

Second Derivatives, Max/Min Points, Inflection Points

1. Find the inflection points for $f(x) = x^4 - 10x^3 + 36x^2 - 27$.

2. If $f(x) = x^3 - 3x$ find 　　　a) $f(2)$ 　　　b) $f'(2)$ 　　c) $f''(2)$.

3. Find the minimum value of y on the graph of $y = 2x^4 - x^2$.

4. a) Does $y = \dfrac{1}{x}$ have an inflection point? Or a max/min point?

   b) Does $y = \dfrac{x+1}{x}$ have an inflection point? Or a max/min point?

   c) Does $y = \dfrac{x^2+1}{x}$ have an inflection point? Or a max/min point?

*5. Find the relative max/min points and inflection points on the graph of

$$y = 3x^4 - 4x^3 - 12x^2.$$

6. If $f(x) = x^{\frac{1}{2}}$ find the following:

   a) $f(4)$ 　　　b) $f'(4)$ 　　c) $f''(4)$

*7. Find the local max/min points and inflection point(s) on the graph of

$$y = x^4 - 8x^3 + 18x^2 - 16x.$$

8. Find derivative of $\dfrac{1}{1+x}$ from first principles.

9. In which quadrants is $\dfrac{dy}{dx}$ positive for $x^2 + y^2 = 25$ (Do not differentiate). In

   which quadrants is $\dfrac{d^2y}{dx^2}$ positive for the same graph?

*10. Given $f'(x) = \dfrac{x^2 - 5x + 4}{x - 5}$

    a) For what values of x is $f(x)$ increasing?

    b) For what values of x is $f'(x)$ increasing?

    c) Does the graph of f change concavity? If so, where?

11. Does $y = x^8$ have an inflection point? If so, where? If not, why not?

12. $f'(x) = \dfrac{x + 1}{x - 2}$

    a) on which intervals is f increasing?

    b) on which intervals is f' increasing?

    c) does f have a local maximum?

Answers to Worksheet 4

1. (2,53) (3,108)   2. a) 2 b) 9 c) 12   3. $-\dfrac{1}{8}$

4. a) No, No   b) No, No   c) No, Rel. Min at (1,2), Rel. Max at (-1,-2)

5. Rel. Min (-1,-5) and (2,-32)   Rel. Max (0,0)

   Inflection points (-0.549,-2.682)   (1.215,-18.35)

6. a) 2   b) $\dfrac{1}{4}$ c) $-\dfrac{1}{32}$ 7. Min point (4,-32), Inflection point (1,-5) (3,-21)

9. Q2 and Q4. Q3 and Q4.

10. a) $1 < x < 4$ or $x > 5$ b) $x < 3$ or $x > 7$ c) at $x = 3$ and $x = 7$

11. No because $\dfrac{d^2 y}{dx^2}$ is always positive.

12. a) $x < -1$ or $x > 2$ b) Nowhere c) Yes, when $x = -1$

Worksheet 5

1. If $f(x) = x\sqrt[3]{x}$ then $f'(x) =$

   (A) $4x^3$ (B) $\dfrac{3}{7}x^{\frac{7}{3}}$ (C) $\dfrac{4}{3}x^{\frac{1}{3}}$ (D) $\dfrac{1}{3}x^{\frac{1}{3}}$ (E) $\dfrac{1}{3}x^{-\frac{2}{3}}$

2. If the line $y = 4x + 3$ is tangent to the curve $y = x^2 + c$, then c is

   (A) 2   (B) 4   (C) 7   (D) 11   (E) 15

3. Suppose that the domain of the function of f is all real numbers and its

   derivative is given by:

   $$f'(x) = \frac{(x-1)(x-4)^3}{1+x^2}$$

   Which of the following is true about the original function of f?

       I. f is decreasing on the interval $(-\infty, 1)$.

       II. f has a local minimum at $x = 4$.

       III. f is concave up at $x = 8$.

   (A) I only   (B) I and II only   (C) I and III only

   (D) II and III only   (E) I, II, III

*4. What are all values of x for which the graph of $y = 6x^2 + \dfrac{x}{2} + 3 + \dfrac{6}{x}$ is concave

   down?

   a) $x < -1$ b) $x < 0$ c) $-1 < x < 0$ d) $0 < x < 1$ e) $x > -1$

graph of f

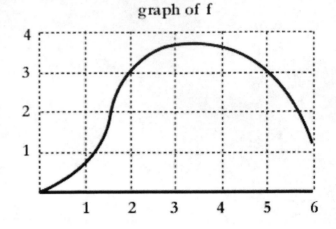

5. A graph of the function f is shown at the right. Which of the following is true?

(A) $\dfrac{f(4)-f(2)}{4-2}=4$

(B) $f(2)>f''(2)$

(C) $f'(2)=f''(2)$ (D) $f'(1)<f'(3)$ (E) None of these

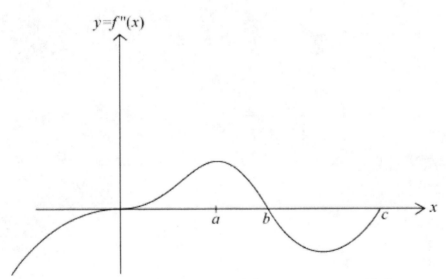

Note: This is the graph of $f''(x)$, **NOT** the graph of $f(x)$.

6. The figure above shows the graph of $f''(x)$, the second derivative of a function $f(x)$. The function $f(x)$ is continuous for all x. Which of the following statements about f is true?

I. f is concave down for $x<0$ and $b<x<c$.

II. f has a relative minimum in the open interval $b<x<c$.

     III. f has points of inflection at $x=0$ and $x=b$.

(A) I only   (B) II only   (C) III only   (D) I and III only   (E) I,II, and III

7. Find the derivative of $y = x^3 + x$ at the point (1,2) using FIRST

PRINCIPLES.

8. For what value of k will $\dfrac{8x+k}{x^2}$ have a relative maximum at $x=4$?

(A) –32   (B) –16   (C) 0   (D)16   (E) 32

9. a) If the graph of $y = x^3 + ax^2 + bx - 8$ has a point of inflection at (2,0), what

    is the value of b?

    b) For which values of x is the function increasing <u>at an increasing rate</u>?

10. The tangent line to the graph of $f(x) = \sqrt[3]{x-2}$ at the point $T(3,1)$ intersects

    the graph of f at another point P.

    a) Write the equation of the line PT.

    b) Find the x - and y - coordinates of point P.

11. The function f is defined on the interval [-4,4] and its graph is shown below.

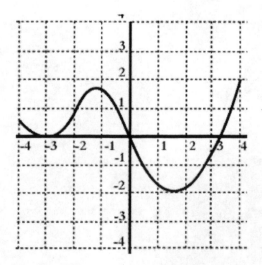

a) Where does f have critical values?

b) on what intervals is f' negative?

c) Where does f' achieve its minimum value? Estimate this value of f'.

d) Sketch a graph of f'.

e) Sketch a graph of f''

Answers to Worksheet 5

1. C

2. C

3. D

4. C

5. B

6. D

8. B

9. a) $+12$ b) $x > 2$

10. a) $y = \dfrac{1}{3}x$ b) $(-6,-2)$

11. a) $x = -3$, -1.4, 1.5

b) $(-4,-3)$ or $(-1.4,1.5)$

c) At $x = 0$. Estimated value of f' is -2.

11. d)

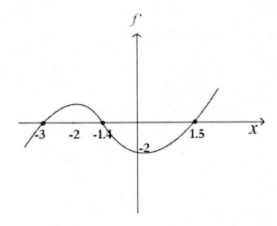

11. e)

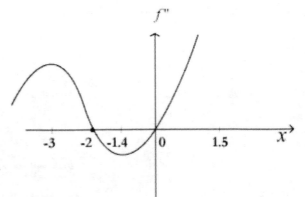

Example

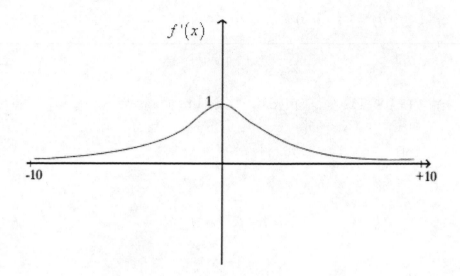

This is a graph of $y = f'(x)$.

The domain of f and f' is $\{x : -10 \le x \le +10\}$.

Question

    a) Does $f(x)$ have a relative maximum or relative minimum point?

Answer

    a) No because $f'(x)$ is never zero.

Question

    b) Does $f(x)$ have an inflection point?

Answer

    b) Yes when $x = 0$ because an inflection point of f occurs at a relative

        maximum point of f'.

Question

    c) If $f(0)=3$ find an approximate x intercept for $y=f(x)$.

Answer

    c) $f'(x)$ is always positive. $\therefore$ f is an increasing function. $f(0)=3$ $\therefore$ a

    graph of $y=f(x)$ looks like:

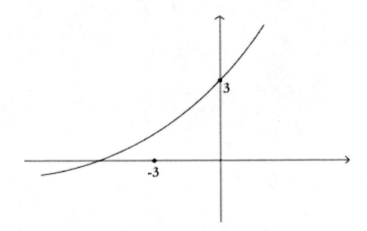

    $f'(x)$ is less than 1 for all x and therefore the slope of any tangent to

    $y=f(x)$ is always less than 1. $\therefore$ x intercept of $y=f(x)$ must be less

    than -3.

Question

    d) How many x intercepts does $y=f(x)$ have?

Answer

    d) Since $y=f(x)$ is always an increasing function it can only have one x

    intercept.

Example

Shown is the graph of the <u>derivative</u> of a function i.e. the graph shown is $y = f'(x)$.

The domain of f is the set $\{x : -3 \le x \le 3\}$.

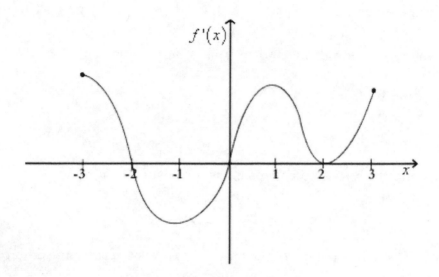

Question

    a) For what values of x does f have a relative minimum?, relative

        maximum?

Answer

    a) When f has a relative minimum point, f' is changing from negative to

    positive.

    $\therefore\ x = 0$ yields a relative minimum point.

    When f has a relative maximum point, f' is changing from positive to

    negative.

    $\therefore\ x = -2$ yields a relative maximum point.

Question

      b) For what values of x is the graph of f concave up?

Answer

      b) The graph of f is concave up when f'' is positive i.e. f' is increasing.

      In the graph shown of $y = f'(x)$, ($-1,1$) and ($2,3$) are two intervals of x

      where f' is increasing and so f is concave up when $-1 < x < 1$ OR

      $2 < x < 3$.

Question

      c) Locate the x co-ordinate of inflection points of f.

Answer

      c) An inflection point of f occurs when f' has a relative maximum or a

      relative minimum value. i.e. $x = -1$ OR $x = 1$ OR $x = 2$

Question

      d) If $f(0) = -1$ sketch a possible graph of $y = f(x)$.

Answer d)

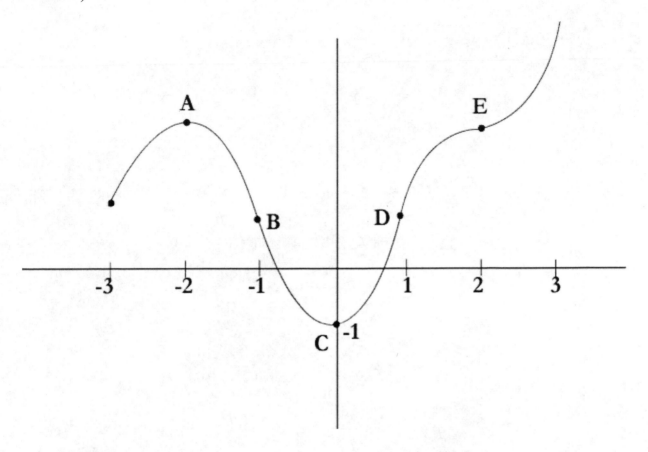

Note: A is a relative maximum point when $x = -2$.

B is an inflection point when $x = -1$

C is a relative minimum point when $x = 0$

D is an inflection point when $x = +1$

E is an inflection point where slope = 0 and when $x = 2$.

Worksheet 6

1. If $f(x) = \sqrt{2x + x^2 + 10}$ then $f'(3)$ equals

   (A) $\dfrac{1}{10}$ (B) $\dfrac{1}{5}$ (C) $\dfrac{4}{5}$ (D) $-\dfrac{4}{5}$ (E) none of these

2. The equation of the tangent line to the curve $y = \dfrac{3x + 4}{4x - 3}$ at the point $(1, 7)$ is

   (A) $y + 25x = 32$ (B) $y - 31x = -24$ (C) $y - 7x = 0$

   (D) $y + 5x = 12$ (E) $y - 25x = -18$

3. Consider the function $f(x) = \dfrac{3x}{k + x^3}$ for which $f'(0) = 1$. The value of k is

   (A) 5   (B) 4   (C) 3   (D) 2   (E) 1

4. The graph of the **first derivative** of a function f is shown at the right. Which of the following are true?

   graph of f'

   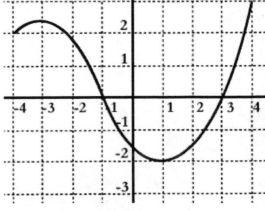

   I.   The graph of f has an inflection point at $x = 1$.

   II.   The graph of f is concave down on the interval $(-3, 1)$.

   III.   The graph of the derivative function f' is increasing at $x = 3$.

   (A) I only   (B) II only   (C) III only   (D) I and II only   E) I, II, III

5. If $f(x) = \dfrac{x-k}{x+k}$ and $f'(0) = 1$, then $k =$

(A) 1   (B) –1   (C) 2   (D) –2   (E) 0

6. The composite function h is defined by $h(x) = f\left[g(x)^2\right]$, where f and g are functions whose graphs are shown below.

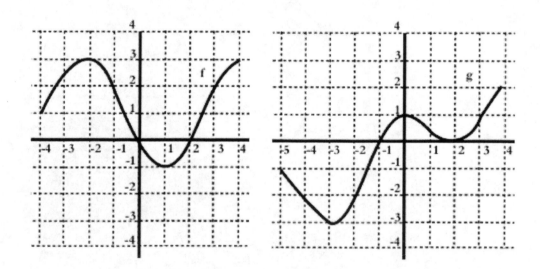

The number of horizontal tangent lines to the graph of h is

(A) 3   (B) 4   (C) 5   (D) 6   (E) 7

7. If $y = \dfrac{2u+1}{2u-5}$ and $u = \sqrt[3]{3x+2} + x$, $\dfrac{dy}{dx}$ at $x = 2$ is:

(A) $\dfrac{-1}{3}$ (B) $\dfrac{-5}{3}$ (C) $\dfrac{-4}{3}$ (D) $\dfrac{5}{4}$ (E) –15

8. The graph of **the second derivative of f** is shown below. Which of the following are true about the function f?

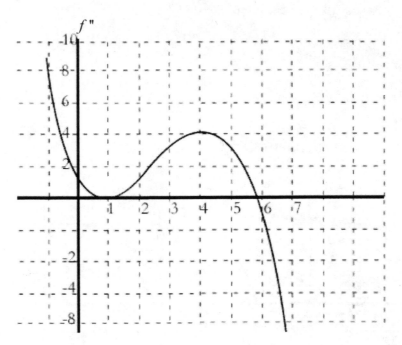

I. f' is decreasing at $x = 0$

II. f is concave up at $x = 5$

III. f has a point of inflection at $x = 1$

(A) I only   (B) II only   (C) I and II only   (D) II and III only   (E) I, II, III

Answers to Worksheet 6

1. C   2. A

3. C   4. D

5. C   6. E

7. B   8. B

Worksheet 7

1. If $f(x) = \dfrac{2x-1}{4x+1}$, then the slope of the tangent at its x intercept is:

    (A) $\dfrac{2}{3}$ (B) 0   (C) undefined   (D) $\dfrac{1}{3}$ (E) $\dfrac{-2}{9}$

2. If $h(x) = f\big(g(2x)\big)$, and it is known that

    $f(1) = -4$, $f'(1) = 5$, $g(6) = 1$, $g'(6) = 9$, then $h'(3)$ equals:

    (A) 45   (B) 30   (C) –15   (D) 90   (E) 60

3. In the graph of $f(x)$ shown below, at which points listed is the slope

    increasing?

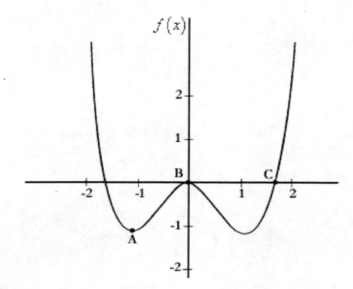

    (A) A only   (B) B only   (C) A, B   (D) A,C   (E) B,C

4. The equation of the tangent to $x + y^2 = 3$ when $y = -1$ is:

   (A) $x - 4y - 2 = 0$ (B) $x + y + 4 = 0$ (C) $x + 4y + 6 = 0$

   (D) $x + 5y + 8 = 0$ (E) $2y - x = -4$

5. $g(x)$ is an **even** function and $f(x)$ is an **odd** function.

   It is known that $f(4) = 6$, $f'(4) = 3$, $g(4) = 8$, and $g'(4) = -3$.

   If $h(x) = \dfrac{f(x)}{g(x)}$, then $h'(-4) =$

   (A) $\dfrac{21}{32}$ (B) $\dfrac{3}{32}$ (C) 21   (D) 32   (E) cannot be found

6. If $y = x - \sqrt{x}$, the minimum value of y is:

   (A) $\dfrac{1}{2}$ (B) $-\dfrac{1}{2}$ (C) 0   (D) $-\dfrac{1}{4}$ (E) -1

7. In the graph of $f(x) = (x-1)^3 (2x-4)^4$, the intervals of x for which the function is decreasing is:

   (A) $x < 1$ only   (B) $\dfrac{10}{7} < x < 2$ only   (C) $x < 1$ or $\dfrac{10}{7} < x < 2$

   (D) $x > 2$ or $x < 1$ (E) $x > 2$ or $1 < x < \dfrac{10}{7}$

8. Consider the graph shown below. State which of the following has the **least** numerical value:

(A) $f(1)$ (B) $f'(1)$ (C) $f''(-1)$ (D) $f(f(1))$ (E) $\dfrac{f(1)-f(0)}{1-0}$

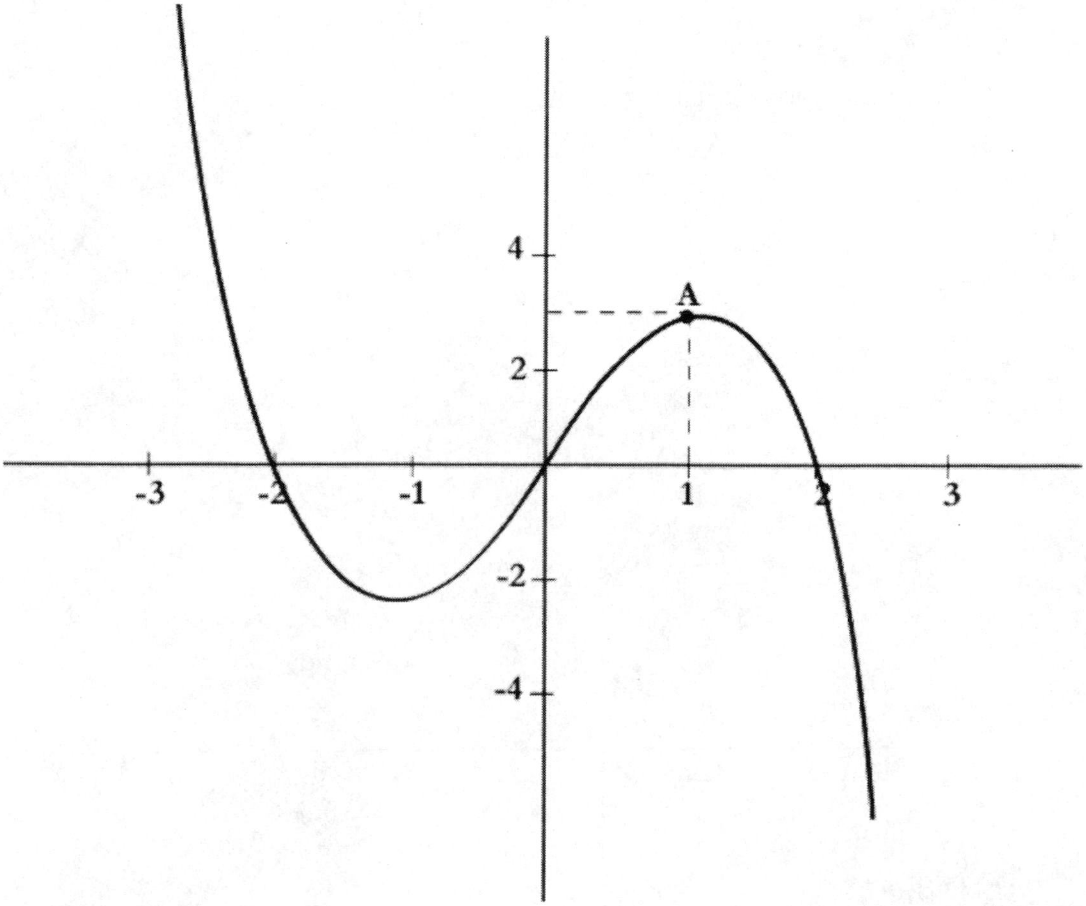

Answers to Worksheet 7

1. A	2. D	3. D	4. E	5. A	6. D
7. B	8. D				

Worksheet 8

1. Let f be the function defined by $f(x) = 3x^5 - 5x^3 + 2$.

   a) On what intervals is f increasing?

   b) On what intervals is the graph of f concave upward?

   c) Write the equation of each horizontal tangent line to the graph.

   d) Sketch an approximate graph of $y = f(x)$.

2.

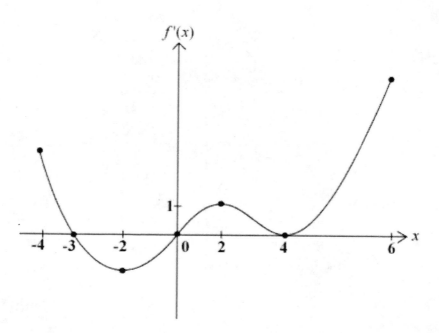

Note: This is a graph of the derivative of f, not the graph of f.

$f'(x)$ is defined for $-4 \le x \le 6$.

The figure above shows the graph of f', the derivative of the function f.

a) For what values of x does the graph of f have a horizontal tangent?

b) For what values of x does f have a relative maximum?

c) For what values of x is the graph of f concave downward?

d) For what values of x does f have an inflection point?

3.

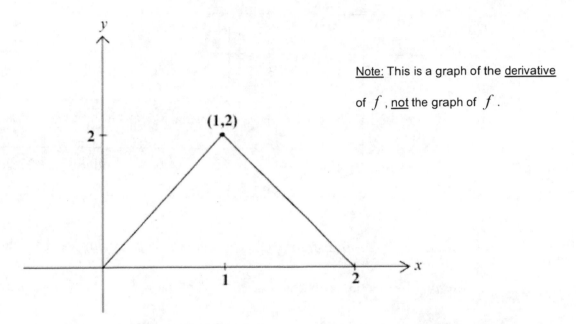

The figure above shows the graph of f', the derivative of a function f. The

domain of f is the set of all x such that $0 \leq x \leq 2$.

a) Does $x = 1$ yield an inflection point on the graph of $y = f(x)$?

b) Is (1,2) a maximum point on the graph of $y = f(x)$?

c) Sketch a graph of $y = f(x)$ given that $f(0) = 1$ and f is a continuous

function.

4. Let f be a function that is <u>even</u> and continuous on the closed interval $[-3,3]$.

   The function f and its derivatives have the properties indicated in the table below.

x	0	$0 < x < 1$	1	$1 < x < 2$	2	$2 < x < 3$
$f(x)$	1	Positive	0	Negative	-1	Negative
$f'(x)$	Undefined	Negative	0	Negative	Undefined	Positive
$f''(x)$	Undefined	Positive	0	Negative	Undefined	Negative

a) Find the x-coordinate of each point at which f attains an absolute maximum value or an absolute minimum value. For each x-coordinate you give, state whether f attains an absolute maximum or an absolute minimum.

b) Find the x-coordinate of each point of inflection on the graph of f. Justify your answer.

c) Sketch the graph of a function with all the given characteristics of f.

Answers to Worksheet 8

1. a) $f'(x) = 15x^4 - 15x^2 = 15x^2(x^2 - 1)$

$f'(x): +$ | $-$ | $-$ | $+$

-1 $\qquad$ 0 $\qquad$ 1 $\qquad$ x

f is increasing when $x \le -1$ or $x \ge 1$.

b) $f''(x) = 60x^3 - 30x = 30x(2x^2 - 1)$

$f''(x) = 0$ when $x = 0$, $\sqrt{\dfrac{1}{2}}$, or $-\sqrt{\dfrac{1}{2}}$

$f''(x)$ $\qquad -$ | $+$ | $-$ | $+$ $\qquad x$

$-\sqrt{\dfrac{1}{2}}$ $\qquad$ 0 $\qquad$ $\sqrt{\dfrac{1}{2}}$

f concave up when $-\sqrt{\frac{1}{2}} < x < 0$ or $x > \sqrt{\frac{1}{2}}$

c) $f'(x) = 0$ when $x = 0$, 1 or -1.

$f(0) = 2 \qquad f(1) = 0 \qquad f(-1) = 4$

Horizontal tangents: $y = 2$, $y = 0$, $y = 4$

d) $x = -2, 2, 4$

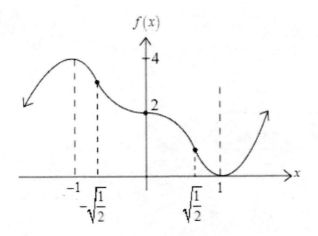

2. a) f has a horizontal tangent at points where $f'(x)=0$.

   This occurs at $x = -3$, 0, and 4.

   b) f has a relative max. at $x = -3$.

   c) f is concave down when $-4 < x < -2$ or $2 < x < 4$.

3. a) Yes

   b) No

   c)

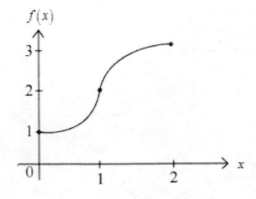

4. a) f is even $\Rightarrow f(x)=f(-x)$, $f'(x)=-f'(x)$; $-1 < f(-3)=f(3) \le 0$

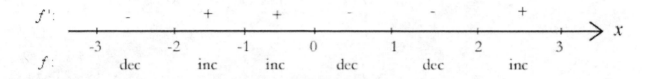

f' :   -   +   +   -   -   +

-3   -2   -1   0   1   2   3   → x

f :   dec   inc   inc   dec   dec   inc

f has absol. max. at $x=0$; absol. min. at $x=\pm 2$

b)

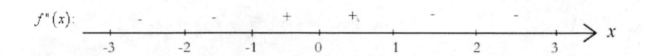

$f''(x)$:   -   -   +   +   -   -

-3   -2   -1   0   1   2   3   → x

f has inflection points at $x=\pm 1$

c)

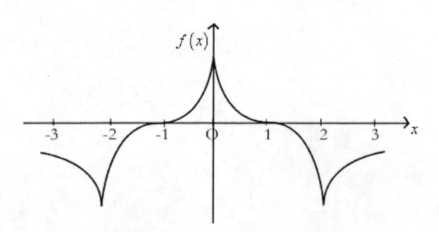

Proof of the Product Rule

Let $y = f(x)g(x)$.

Then $\dfrac{dy}{dx} =$ (by definition) $\displaystyle\lim_{h\to 0} \frac{f(x+h)g(x+h) - f(x)g(x)}{h}$

$$= \lim_{h\to 0} \frac{f(x)g(x+h) - f(x)g(x) + f(x+h)g(x+h) - f(x)g(x+h)}{h}$$

$$= \lim_{h\to 0} \frac{f(x)\left[g(x+h) - g(x)\right]}{h} + \lim_{h\to 0} g(x+h)\frac{\left[f(x+h) - f(x)\right]}{h}$$

$$= f(x)\lim_{h\to 0}\frac{g(x+h) - g(x)}{h} + \lim_{h\to 0} g(x+h)\lim_{h\to 0}\frac{f(x+h) - f(x)}{h}$$

$$= f(x)g'(x) + g(x)f'(x)$$

Proof of the Quotient Rule

Let $y = \dfrac{f(x)}{g(x)} = f(x)\cdot\left[g(x)\right]^{-1}$

By the Product Rule and Chain Rule:

$$\frac{dy}{dx} = \left[g(x)\right]^{-1} f'(x) + f(x)(-1)\left[g(x)\right]^{-2}\cdot g'(x)$$

$$= \frac{f'(x)}{g(x)} - \frac{f(x)g'(x)}{\left[g(x)\right]^{2}}$$

$$= \frac{g(x)f'(x) - f(x)g'(x)}{\left[g(x)\right]^{2}}$$

Proof of the Chain Rule

The Chain Rule is best shown by using dy and dx notation (attributable to Liebnitz).

Suppose $y = f(x) = g(h(x))$ then this can be decomposed as: $f(x) = g(u)$

where $u = h(x)$.

i.e. $\left. \begin{array}{l} y = g(u) \\ u = h(x) \end{array} \right\}$

We wish to find $\dfrac{dy}{dx}$.

We wish to show $\dfrac{dy}{dx} = \dfrac{dy}{du} \div \dfrac{dx}{du}$.

$\dfrac{dy}{dx} = (\text{by definition}) \quad \lim\limits_{\Delta x \to 0} \dfrac{\Delta y}{\Delta x} = \lim\limits_{\Delta x \to 0} \left(\dfrac{\Delta y}{\Delta u} \cdot \dfrac{\Delta u}{\Delta x} \right)$

$$= \lim\limits_{\Delta x \to 0} \dfrac{\Delta y}{\Delta u} \cdot \lim\limits_{\Delta x \to 0} \dfrac{\Delta u}{\Delta x}$$

Since u is a continuous function of x, it follows that as $\Delta x \to 0$ then $\Delta u \to 0$.

$$\therefore \dfrac{dy}{dx} = \lim\limits_{\Delta u \to 0} \dfrac{\Delta y}{\Delta u} \cdot \lim\limits_{\Delta x \to 0} \dfrac{\Delta u}{\Delta x}$$

$$= \dfrac{dy}{du} \cdot \dfrac{du}{dx} = \dfrac{dy}{du} \div \dfrac{dx}{du}$$

CHAPTER 3

Graphing Polynomials

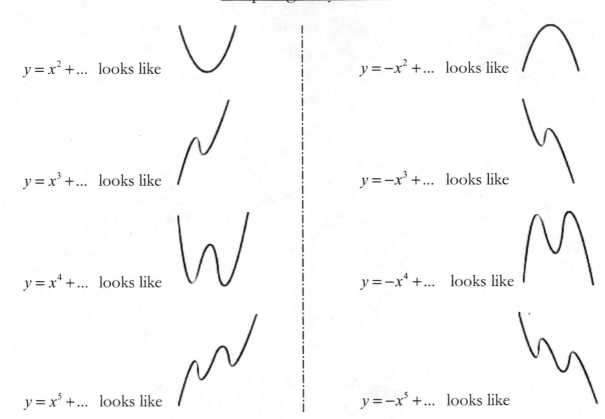

$y = x^2 + \dots$ looks like

$y = -x^2 + \dots$ looks like

$y = x^3 + \dots$ looks like

$y = -x^3 + \dots$ looks like

$y = x^4 + \dots$ looks like

$y = -x^4 + \dots$ looks like

$y = x^5 + \dots$ looks like

$y = -x^5 + \dots$ looks like

Usually the graph of a polynomial will have one fewer "bends" than the degree of the polynomial. If we consider each graph as a piece of string which can be "pulled" then note that we can distort the shapes so that, for example, a cubic polynomial may look like :

Or, a quartic polynomial may look like

From the point of view of graphing a polynomial, the most important characteristics

of the polynomial graph are its intercepts.

For example, $y = (x-1)(x-2)(x-3)$ has intercepts of (1,0) (2,0) (3,0) and (0,-6)

and it is easy to deduce that its general shape is

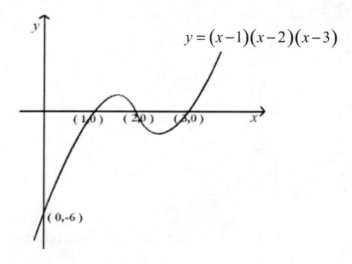

Note that in graphing $f(x) = (x-2)^2$ the curve "bounces" at (2,0) since $f(2^-)$ is

positive and $f(2^+)$ is also positive i.e. f does not change sign as it passes

through (2,0).

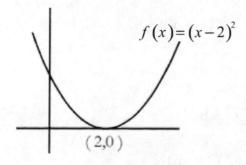

A similar situation obtains in the graph of (say)

$$f(x)=(x-1)(x-2)^2(x-3)$$

since $f(2^-)$ is negative and $f(2^+)$ is negative also.

Graph looks like:

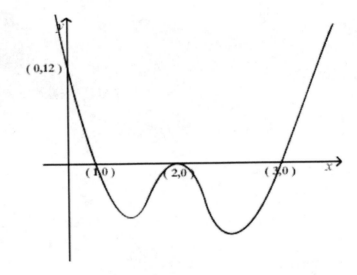

Remember that in the following worksheet you should check your graphs by using your graphing calculator.

Worksheet 1

1. Sketch the following:

   a) $y=x^2$ b) $y=-(x-1)(x-2)$

   c) $y=x^3$ d) $y=(x-1)(x-2)(x+2)$

2. Sketch the following:

   a) $y=(x-1)(x-2)^2(x-3)$ b) $y=(x-1)(x-2)^3(x-3)$

   c) $y=(x+1)^3(x+2)^2(x+5)$

3. Graph the following, showing intercept(s), max/min points, inflection points.

   a) $y = x^4 - 2x^2$ b) $y = 2x^3 - 3x^2$

*4. Graph the following, showing intercept(s) and critical points.

   a) $y = x^4 - 4x^3$ b) $y = -x^3 + 12x + 16$

   c) $y = x^3 - 6x^2 + 9x - 2$ d) $y = x^3 - 6x^2 + 9x - 4$

   e) $y = x^3 + 3x^2 - 24x + 28$

*5. Graph $y = x^3 - 12x + 20$ showing max/min points and inflection points. Show

   an approximate value for the x intercept(s).

6. Find points on $y = x^3 - 2x$ where the tangent makes an angle of $45°$

   with the x axis.

7. Graph $y = x^4 - 2x^3$ for $-2 \le x \le 4$.

8. Graph $y = x^3 - 3x^2$. Use this graph to help you sketch $y = x^3 - 3x^2 + 2$.

9. Find the equations of <u>two</u> tangents to $y = x^2$ which pass through (1,0).

   Repeat so that the tangents pass through (2,3).

10. Explain why it is true that the tangent to any cubic polynomial at any point

   always intersects the curve again, except for a single point on the cubic

   polynomial graph. Identify that point.

11. Prove that every cubic polynomial function has a point of symmetry

   i.e. prove that there exists a point M on the graph so that for every point P on

   the graph there exists a point Q on the graph so that M is the mid-point of

   PQ.

Answers to Worksheet 1

3. a) Min (1,-1) (-1,-1)

   Max (0,0)

   Intercepts ($\sqrt{2}$,0) ($-\sqrt{2}$,0) (0,0)

   Inflection Points ($\frac{1}{\sqrt{3}}$, $-\frac{5}{9}$) ($-\frac{1}{\sqrt{3}}$, $-\frac{5}{9}$)

   b) Min (1,-1)   Max (0,0)

   Inflection ($\frac{1}{2}$, $-\frac{1}{2}$)

   Intercepts (0,0) ($1\frac{1}{2}$,0)

4. a) Min (3,-27)   Inflection (0,0) (2,-16)

   Intercepts (0,0) (4,0)

   b) Min (-2,0)   Max (2,32)

   Inflection (0,16)

   Intercepts (-2,0) (4,0)

   c) Max (1,2)   Min (3,-2)

   Inflection (2,0)

   Intercepts ($2-\sqrt{3}$,0) ($2+\sqrt{3}$,0) (2,0)

   d) Max (1,0) Min (3,-4)   4. e) Max (-4,108)   Min (3,-4)

   Inflection (2,-2)   Inflection (-1,54)

   Intercepts (1,0) (0,-4) (4,0)   Intercepts (-7,0) (2,0) (0,28)

5. Max (-2,36)   Min (2,4)

Inflection (0,20)

Intercepts (0,20)   (-4.107,0)

6. (1,-1) and (-1,1)

7. Min $\left(\frac{1}{2}, -\frac{27}{16} \right)$ Inflections (0,0)   (1,-1)

Intercepts (0,0)   (2,0)

8. Max (0,0)   Min (2,-4)

Inflection (1,-2)

Intercepts (0,0) and (3,0)

9. i) $y = 0$ and $y = 4x - 4$

ii) $y = 2x - 1$ and $y = 6x - 9$

10. The inflection point of the graph

11. M is the inflection point of the graph

As we saw in Chapter 2, the relative minimum point P of a graph means a point whose y co-ordinate is less than the y co-ordinates of those points in the immediate neighbourhood of P.

This does not mean, of course, that P is necessarily the lowest point on the whole curve as seen below.

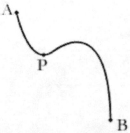

We say P is a relative minimum point and B is an absolute minimum point on the curve. The absolute minimum point on a curve means the lowest point on the curve and may well occur at a point where the slope is <u>not</u> zero or possibly not even defined.

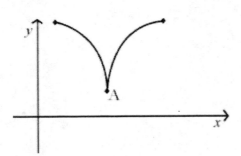 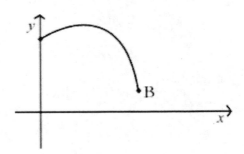

Both A and B are absolute minima. These types of situations often occur where the domain of the function is restricted to a closed set.

Example 1

Question: Find:

a) the relative minimum   b) the absolute minimum

c) the relative maximum   d) the absolute maximum

for the function $f(x) = 2x^3 - 9x^2 + 12x$ restricted to the

domain $\{x : 0 \le x \le 3\}$

Answer:

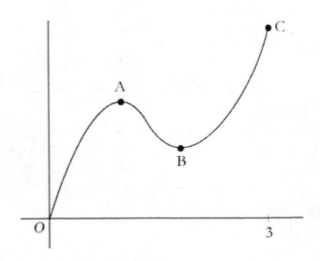

Since $f(x)$ is a polynomial we know that the graph is continuous and has no cusps. By differentiating and solving $f'(x)=0$ can determine that A is ($1,5$) and B is ($2,4$).

Note that the end-points O is ($0,0$) and C is ($3,9$).

It follows that the relative minimum point is ($2,4$).

the absolute minimum point is ($0,0$).

the relative maximum point is ($1,5$).

the absolute maximum point is ($3,9$).

Worksheet 2

1. Sketch $y=(x+1)^3(x-2)^2(x-4)$. You do not need to show max/min points.

2. On the graph shown below name the points where

    a) $x=0$ b) $f(x)=0$ c) $f'(x)=0$

    d) $f''(x)=0$ e) $f'(x)=f(x)$

    f) $f''(x)>0$ and $f'(x)>0$

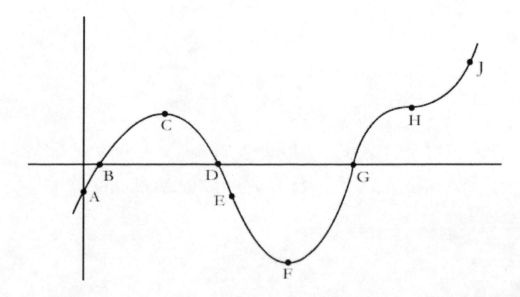

3. Explain in your own words why no tangent to $y = 2x^3 - x$ is parallel to $y = 4 - 2x$.

4. For $y = x^3 - x^2 - 4x + 4$ find the equation of the tangent at the point (2,0).

   Find the x co-ordinate of another point on this curve so that the tangent at this second point is parallel to the tangent at (2,0).

5. On the graph shown below state the points where

   a) $x = 0$ b) $f(x) = 0$ c) $f'(x) = 0$

   d) $f''(x) = 0$ e) $f(x)$ is increasing

   f) $f'(x)$ is increasing

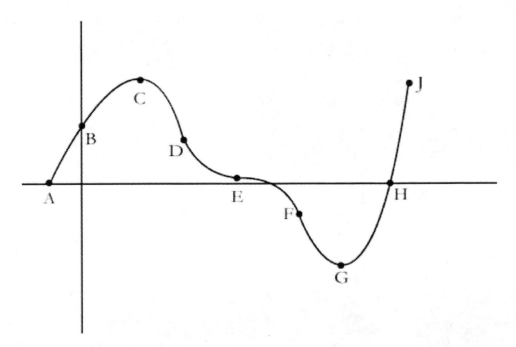

6. Find   a) relative max   b) relative min   c) absolute max

   d) absolute min for $y = 2x^3 - 9x^2 + 12x$ if x is restricted to the

   domain $\{x \mid 0 \le x \le 3\}$

<u>Answers to Worksheet 2</u>

2. a) A   b) B, D, G   c) C, F, H   d) E, G, H   e) E, J possibly

 f) J

4. $y = 4x - 8$

5. a) B   b) A, H   c) C, E, G   d) D, E, F

 e) A, B, G, H   f) D, F, G, H

6. a) $(1,5)$ b) $(2,4)$ c) $(3,9)$ d) $(0,0)$

Worksheet 3

1. Name the algebraic expression which is equivalent to

 $$\lim_{h \to 0} \frac{(x+h)^5 + 2(x+h) - (x^5 + 2x)}{h}$$ No explanation required.

2. Find   a) relative max   b) relative min

 c) absolute max   d) absolute min

 for the graph of $y = x^3 - 6x^2 + 9x - 4$ if x is restricted to the

 domain $\{x \mid -1 \le x \le 5\}$.

3. Sketch a graph of $y = f(x)$ for which the following table is valid.

x	$f(x)$	$f'(x)$	$f''(x)$
0	−	−	+
1	−	0	+
2	0	+	0
3	+	+	−
4	+	0	−

4. For which values of x does $y = (x-3)^{10}(x^2+1)^5$ have a relative minimum?

5. Find the equations of the tangents, with slope equal to 9, to the graph of

   $y = x^2(x-3)$.

6. Find the derivative of $f(x) = (1-x)^{-1}$ by first principles.

7. Graph $y = x^3 + 3x^2 - 24x + 28$. Show intercept(s), relative max/min points and

   inflections.

8. Use your graphing calculator to find the following:

   a) The positive root nearest $x = 0$ of the equation $x^4 - 2^x - 9 = 0$ (to two

      decimal places).

   b) The minimum value of the function $y = x^4 - 2^x - 9$ in the interval $[0,6]$

      (to two decimals).

   c) The maximum value of the function $y = x^4 - 2^x - 9$ in the interval $[10,18]$.

9. Let f and g be differentiable functions such that

   $f(1) = 2 \qquad f'(1) = 3 \qquad f'(2) = -4$

   $g(1) = 2 \qquad g'(1) = -3 \qquad g'(2) = 5$

   If $h(x) = f(g(x))$, then $h'(1) =$

   (A) –9 \qquad (B) –4 \qquad (C) 0 \qquad (D) 12 \qquad (E) 15

10. Let the function f be differentiable on the interval $[\,0,2.5\,]$ and define g by

$g(x)=f\bigl(f(x)\bigr)$. Use the table to estimate $g'(1)$.

x	0.0	0.5	1.0	1.5	2.0	2.5
$f(x)$	1.7	1.8	2.0	2.4	3.1	4.4

(A) 0.8   (B) 1.2   (C) 1.6   (D) 2.0   (E) 2.4

Answers to Worksheet 3

1. $5x^4+2$

2. a) $(\,1,0\,)$ b) $(\,3,-4\,)$ c) $(\,5,16\,)$ d) $(\,-1,-20\,)$ 4. $x=3$ or $\dfrac{1}{2}$

3.

5. $y=9x-27$ and $y=9x+5$

8. a) $x=1.89$ b) -10.39

     c) 22036.41

9. D

10. B

Worksheet 4

1. The domain of an even function f is all real numbers. If a tangent line to the

graph of f at $x=4$ has slope -2 and crosses the x axis at $x=10$, which of

the following is true?

         I. $f(4)=12$

         II. $f'(4)=-2$

         III. $f'(-4)=2$

(A) II only   (B) I and II only   (C) I and III only

(D) II and III only   (E) I, II, III

2. Suppose that the domain of the function f is all real numbers and its derivative is given by

$$f'(x)=\frac{(x-2)(x-3)^3}{1+x^2}$$

Which of the following is true about the original function f?

      I. f is increasing on the interval $(-\infty,2)$

      II. f has a local minimum at $x=3$.

      III. f has a local maximum at $x=2$.

(A) I only   (B) I and II only   (C) I and III only

(D) II and III only   (E) I, II, III

3. At which of the three labeled points on the graph of f in the figure below is it possible for $f'(x)=f''(x)$?

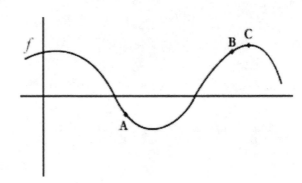

(A) A only   (B) B only   (C) C only (D) A, B and C   (E) none of A, B, C

4. The graph of the **derivative of f** is shown.

    Which of the following is true about the

    function f?

    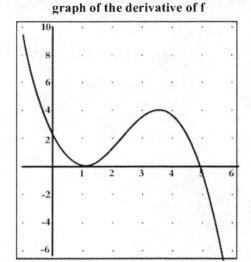

    graph of the derivative of f

    I. f is increasing at $x = 0$.

    II. f has a local minimum at $x = 1$.

    III. f is concave up at $x = 3$.

    IV. f has a local maximum at $x = 5$.

    (A) All of them   (B) I, III, IV only   (C) II only

    (D) I and III only   (E) II, III only

5. The graph of the **second derivative** of a function f is shown at the right.

    Which of the following is true?

    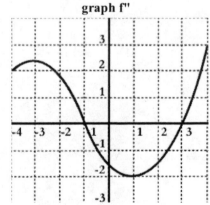

    graph f''

    I.   The graph of f has an inflection point at

           $x = -3$.

    II.   The graph of f is concave down on the

           interval (-1,3).

    III.   The graph of the derivative function f' is increasing at $x = -3$.

    (A) I only   (B) II only   (C) III only   (D) I and II only   (E) I, II, III

6. Suppose a function f is defined so that it has derivatives

    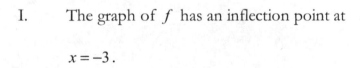
    $f'(x) = x^2 (x-2)$ and $f''(x) = x^2 - x$.

    Over what interval is the graph of f both increasing and concave up?

    (A) $x < 0$ (B) $x < 2$ (C) $x > 2$ (D) $0 < x < 2$ (E) $x > 1$

7. The graph of g is shown below.

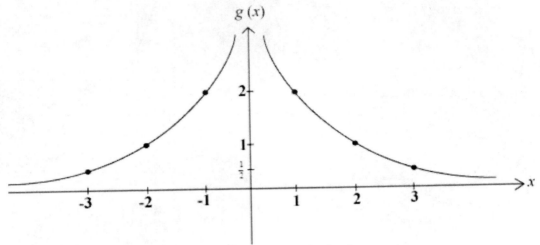

$g(x)$

Use the graph to say which of the following is (are) true:

I. $\displaystyle\lim_{h\to 0}\frac{g'(2+h)-g'(2)}{h}$ is positive.

II. $g(g(2))=2$

III. $g(x)=g^{-1}(x)$

IV. $g'(-3)=\dfrac{1}{g'(3)}$

A) II only   B) I, III only   C) II, III only   D) IV only   E) I, II only

8. If $f(x)=\dfrac{x^2+x+1}{x+1}$, then $f'(0)=$

(A) 0   (B) 1   (C) 2   (D) 3   (E) none of these

9. Let f and g be differentiable functions such that

$$f(1)=2 \qquad f'(1)=1 \qquad f'(2)=-1$$

$$g(1)=2 \qquad g'(1)=2 \qquad g'(2)=1$$

If $h(x)=f\big(g(x)\big)$, then $h'(1)=$

(A) 1   (B) 2   (C) 4   (D) –2   (E) none of these

10. Let the function f be differentiable on the interval [0, 2.5] and

define g by $g(x)=f\big(f(x)\big)-f(x)$.

Use the table to estimate $g'(1)$.

x	0.0	0.5	1.0	1.5	2.0	2.5
$f(x)$	1.7	1.8	2.0	2.4	3.1	4.4

(A) 0.8   (B) 0.6   (C) 0.4   (D) 1   (E) 1.2

11. The line $y=4x+k$ is tangent to the curve $y=x^4$ when k is equal to

(A) –3 only   (B) 0 only   (C) –3 or 5   (D) –2 or 4   (E) –1 or 6

12. The composite function h is defined by $h(x)=f\big[g(x)\big]$, where f and g are

functions whose graphs are shown below.

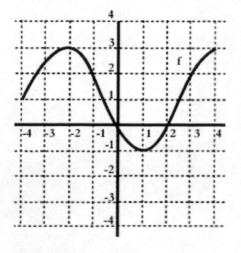
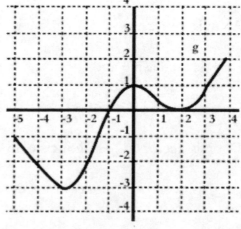

The number of relative minimum points on the graph of h is

(A) 0   (B) 1   (C) 2   (D) 3   (E) 4

13. The slope of a curve at any point (x, y) is $\dfrac{dy}{dx} = (x-2)^2 (x-3)$. At the point where $x = 2$ the curve has a

     I.   an x-intercept.

     II.   local maximum.

     III.   point of inflection.

(A) I only   (B) II only   (C) III only   (D) I and II only   (E) I and III only

14. Let $f(x) = x^5 - 3x^2 + 4$. For how many inputs c between $a = -2$ and $b = 2$ is it true that $\dfrac{f(b) - f(a)}{b - a} = f'(c)$?

(A) 0   (B) 1   (C) 2   (D) 3   (E) 4

15. The table below has some values of the derivative of a continuous function g.

x	1.94	1.96	1.98	2.00	2.02	2.04
$g'(x)$	3.92	3.97	3.99	4.00	3.99	3.96

Based on this information it appears that on the interval covered by the table

(A) g is increasing and concave up everywhere.

(B) g is increasing and concave down everywhere.

(C) g has a point of inflection

(D) g is decreasing and concave up everywhere.

(E) g is decreasing and concave down everywhere.

16. Which of the following statements is *false?*

   (A) If c is a critical number of the function f, then it is also a critical number

   of the function $g(x)=f(x)+k$, where k is a constant.

   (B) If a function f is continuous on a closed interval then it must have a

   minimum value on the interval.

   (C) If a function has 3 zeros then it must have at least 2 points where the

   tangent line is horizontal.

   (D) The maximum value of a function that is continuous on a closed interval

   can occur at two different inputs on the graph.

   (E) The graph of a function can have at most two horizontal asymptotes.

17. Suppose f is a continuous and differentiable function on the interval $[0,1]$

   and $g(x)=x-f(2x)$. The table below gives some values of f.

x	0.1	0.2	0.3	0.4	0.5	0.6
$f(x)$	1.010	1.042	1.180	1.298	1.486	1.573

   What is the approximate value of $g'(0.1)$?

   (A) 0.15   (B) –0.7   (C) –1.4   (D) 0.3   (E) 0

18. The composite function h is defined by $h(x) = f\big[g(x)\big]$, where f and g are

functions whose graphs are shown below.

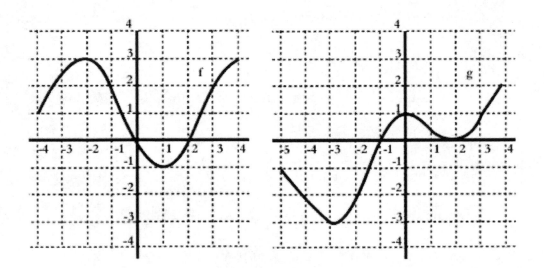

h has a relative maximum point when $x =$

(A) -3   (B) -2   (C) 0   (D) 1   (E) 3

Answers to Worksheet 4

1. E   2. E   3. E   4. B   5. E   6. C   7. A   8. A   9. D

10. B   11. A   12. D   13. C   14. C   15. C   16. C   17. B   18. B

Rational Functions

We will define a rational function to be a function of the form $y = h(x) = \dfrac{f(x)}{g(x)}$

where f and g are polynomial functions.

When graphing rational functions the use of a graphing calculator is of course very helpful, but it is important to understand how the characteristics of the equation relate to the graph.

Notice for example that since division by zero leads to "difficulties" it is clear that when $g(x) = 0$ the graph of $y = h(x)$ will have a discontinuity. It is recommended that an analysis of the graph of a rational function should focus on the following in the order

1) intercepts

2) asymptotes

3) relative max/min points

An asymptote is a straight line to which the curve "gets very close" as x or y (or both) gets large. The curve may cross the asymptote in some cases.

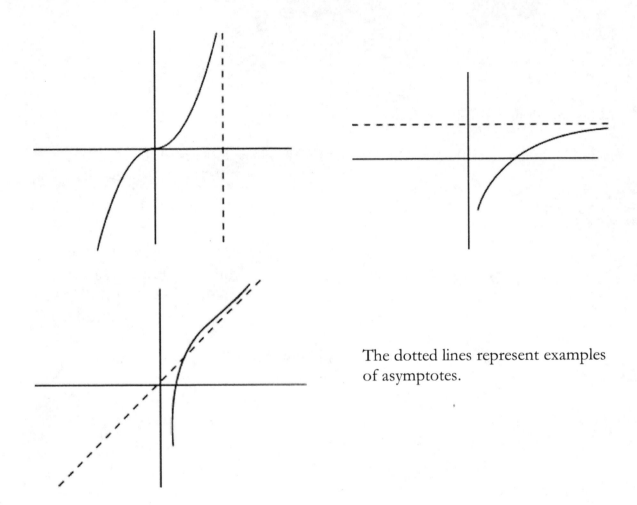

The dotted lines represent examples of asymptotes.

It cannot be emphasized too strongly that when asymptotes exist then x or y (or both) become large.

Remember that $\dfrac{1}{.001}=1000$ and consequently it is a good idea to think intuitively of

$\dfrac{1}{0}$ as a <u>big</u> number (rather than undefined). Similarly $\dfrac{1}{big}$ should be considered as

'equal' to zero.

For these reasons, for the graph of

$$y = 2 + \frac{1}{(x-1)^2}$$

a) When x "equals" 1, y is big

   We say $x = 1$ is an asymptote since as x gets close to 1, y gets very large.

b) When x gets big, y gets close to 2

   We say $y = 2$ is an asymptote.

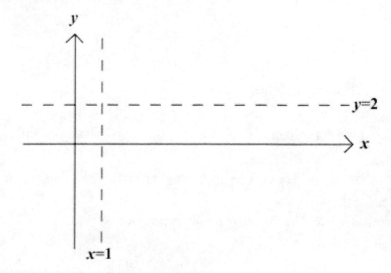

For the foregoing reasons when x or y get large the curve gets close to the dotted lines shown. We know that the only intercept is (0,3) and hence it is easy to deduce that the graph looks something like

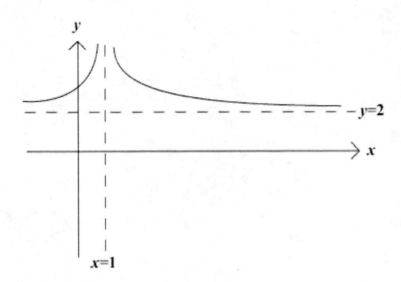

Note that when $x = 1^-$, y is $+\infty$.

and when $x = 1^+$, y is $+\infty$ also.

i.e. the sign of y does not change as x "passes through" the value of 1. This is a result of the fact that the power of $(x-1)$ is an <u>even</u> integer and explains why the graph is asymptotic to $x = 1$ in the positive y sense only.

Example

$$y = h(x) = \frac{3x-6}{x-1}$$

1) Intercepts are (2,0) and (0,6)

2) Asymptotes occur when the denominator equals zero since then y will be

large.

i.e. $x = 1$ is an asymptote

Also, when x is large, y is close to 3

i.e. as x "approaches" infinity, y approaches 3.

These two characteristics help us to deduce that the graph looks like

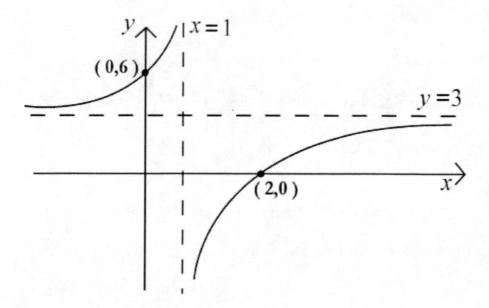

It seems clear that there are no relative max/min points, which can be verified

by differentiating.

Example

To graph $y = f(x) = \dfrac{x^2}{x-1} = x+1+\dfrac{1}{x-1}$:

1) The only intercept is $(0,0)$.

2) As x approaches 1^-, y approaches $-\infty$.

   As x approaches 1^+, y approaches $+\infty$.

   i.e. $x=1$ is an asymptote.

   Also as x approaches $+\infty$ or $-\infty$, y approaches $x+1$.

   i.e. $y=x+1$ is an asymptote.

3) Relative max/min points

   $$f'(x) = 1 - \frac{1}{(x-1)^2}$$

   When $f'(x)=0$, $x=2$ or 0.

   i.e. Potential relative max/min points occur at $(2,4)$ and $(0,0)$.

   Since $f''(x) = \dfrac{2}{(x-1)^3}$ then $f''(2)>0$ and $f''(0)<0$.

   $\therefore$ $(2,4)$ is a relative minimum point and

   $(0,0)$ is a relative maximum point.

Hence the graph looks like:

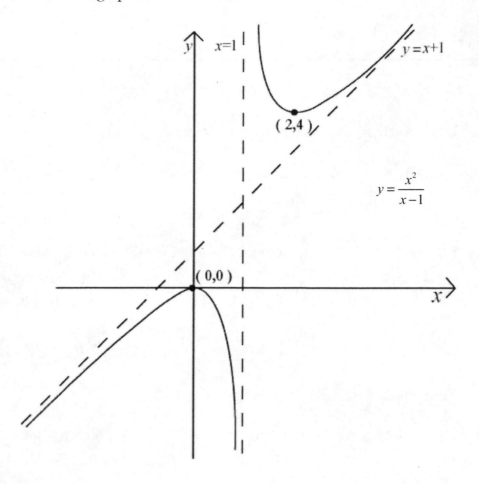

A crude and incomplete, but nevertheless helpful, method for locating asymptotes is

1) make denominator equal to zero

2) divide denominator into numerator and discard the remainder

We are reminded once again that when an asymptote is in evidence then x or y (or

both) become large.

For example, $y = \sqrt{x^2 - 2x}$ has an asymptote of $y = x - 1$ since as $x \to +\infty$ then

$\sqrt{x^2 - 2x}$ approximates to $\sqrt{x^2 - 2x + 1} = x - 1$.

Similarly as $x \to -\infty$, $y \to 1 - x$ and hence $y = 1 - x$ is also an asymptote.

Worksheet 5

1. Let $f(x) = \dfrac{x+1}{x^2}$.

   a) Find the x intercept(s).

   b) Explain why the graph has no y intercepts.

   c) Find where the slope is zero. Is this a max or a min?

   d) Find the equation of the tangent to the curve at $(-1,0)$.

   e) Find the equation of the straight line to which the curve approximates as x gets very large.

2. For the function $y = \dfrac{x^2+4}{x}$

   a) Find where the tangent is parallel to the x axis.

   b) Decide which points are relative max and which are relative min.

   c) Locate the asymptotes.

   d) Find the minimum value of $\dfrac{x^2+4}{x}$ assuming x is positive.

   e) Try to draw a neat large graph of the function.

3. Graph the following functions indicating intercept(s), max/min points, and asymptotes where appropriate.

   a) $y = \dfrac{x+1}{x}$ b) $y = \dfrac{x-1}{x}$ c) $y = \dfrac{x^2-x+1}{x-1}$ d) $y = \dfrac{x^3-1}{x^2}$

   e) $y = x + \dfrac{1}{x}$ f) $y = x + \dfrac{1}{x^2}$ g) $y = \dfrac{x^3-4x^2+4x}{x^2-2x+1}$

Answers to Worksheet 5

1. a) $(-1, 0)$

    b) x cannot equal zero.

    c) $(-2, -\frac{1}{4})$ is a minimum point

    d) $y = x + 1$

    e) $y = 0$

2. a) $(2, 4)$ $(-2, -4)$

    b) $(2, 4)$ is a min, $(-2, -4)$ is a max.

    c) $y = x$ and $x = 0$ d) 4

    e)

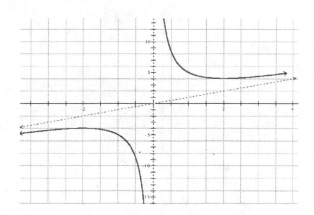

3. a)

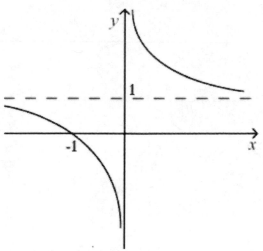

3. b)

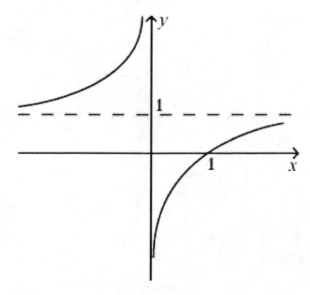

3 c)

Min $(2, 3)$
Max $(0, -1)$

3. d)

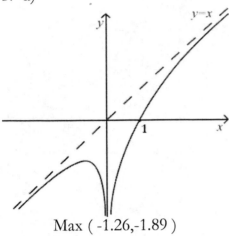

Max (-1.26,-1.89)

3. e)

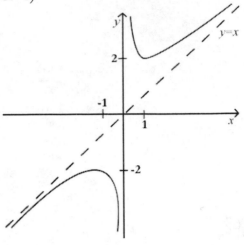

Min (1,2)
Max (-1,-2)

3. f)

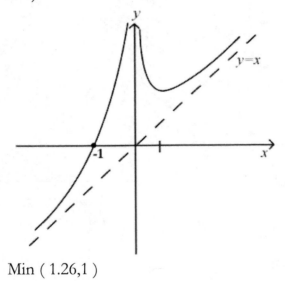

Min (1.26,1)

3. g)

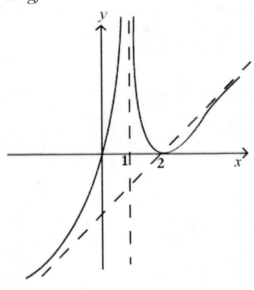

Min (2,0)

Worksheet 6

1. State a possible equation for the function graphed below.

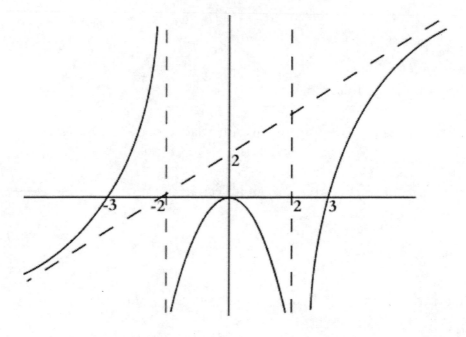

2. State a possible equation for the function graphed below.

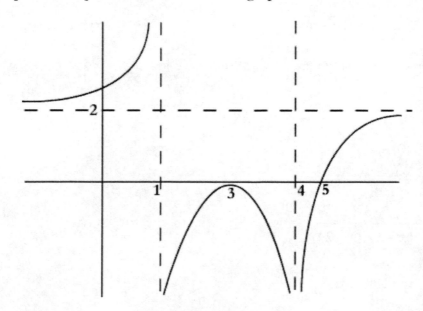

3. Solve the following inequality $-3 \leq \dfrac{(x-1)(x-5)}{x-3} \leq 3$.

Hint: Use a graph of $y = \dfrac{(x-1)(x-5)}{x-3}$.

4. Find the set of values of y for which x is real when :

$$y = \frac{15 + 10x}{4 + x^2}$$

5. Investigate the following:

"If $f(x)$ is a cubic polynomial with relative max. at $(a, f(a))$ and relative

min. at $(b, f(b))$ then f has an inflection point at $\left(\frac{a+b}{2}, \frac{f(a)+f(b)}{2} \right)$.

6. Find the equation(s) of the asymptote(s) to $y = \sqrt{x^2 - 2x} + \frac{x^2 + 1}{x + 1}$.

7. Use your calculator to estimate the value of the following expressions as x

approaches positive infinity and negative infinity.

a) $x \sin \frac{1}{x}$ b) $\frac{2^x - 1}{2^{-x} - 1}$ c) $\sqrt{x^2 + 2x} - \sqrt{x^2 - 2x}$

Answers to Worksheet 6

1. $y = \frac{x^2 (x+3)(x-3)}{(x-2)^2 (x+2)}$ 5. Conjecture is true

2. $y = \frac{2(x-3)^2 (x-5)}{(x-4)^2 (x-1)}$ 6. $y = 2x - 2$, $y = 0$

3. $-1 \leq x \leq 2$ OR $4 \leq x \leq 7$ 7. a) 1, 1   b) $-\infty$, 0   c) 2, -2

4. $-\frac{5}{4} \leq y \leq 5$

Worksheet 7

1. Shown is the graph of the derivative of the function f. i.e. The graph shown is $y = f'(x)$. The domain of f is the set $\{x \mid -3 \le x \le 3\}$.

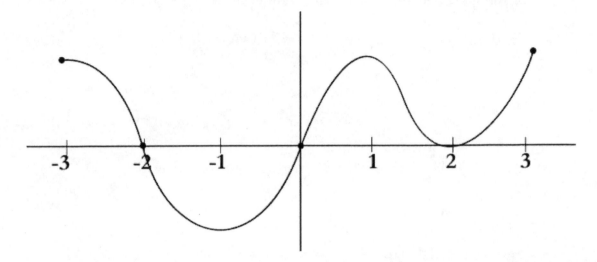

a) For what values of x between -3 and 3 does f have a relative maximum, a relative minimum?

b) For what values of x between -3 and 3 is the graph of $y = f(x)$ concave up?

c) Sketch a possible graph of $y = f(x)$.

2. State a possible equation for the graph sketched below.

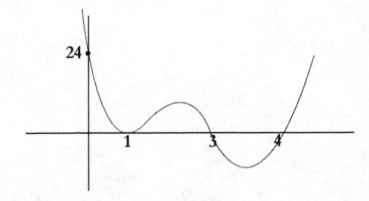

3. State an equation of the asymptote(s) of the following graph:

$$y = \frac{x^3 + 1}{x^2 - 2x}$$

Without sketching the graph and without the aid of a graphing calculator find

the point(s) where the graph crosses its own asymptote(s).

4. If $f'(x)$ and $g'(x)$ are equal for all values of x does it mean that

$f''(x) = g''(x)$ for all values of x? Is the converse true?

5. The point (-2,0) lies on the graph of $y = x^4 + ax^3 + cx^2$.

Also this point, (-2,0), is a point of inflection. Find the value(s) of a and c.

6. Graph, without the aid of a graphing calculator:

$$y = \frac{x}{\sqrt{x^2 - 4}}$$

State the equations for the horizontal and vertical asymptotes.

7. The tangent to $y = \sqrt{x}$ at P (9,3) intersects the x axis at Q. Prove that the y

axis bisects PQ.

Answers to Worksheet 7

1. a) -2 rel max. 0 rel min.

   b) $-1 < x < 1$ or $x > 2$

   c)

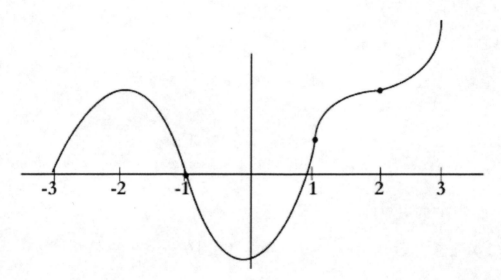

2. $y = 2(x-1)^2 (x-3)(x-4)$

3. $y = x+2$, $x = 0$, $x = 2$ $\qquad \left(-\dfrac{1}{4}, 1\dfrac{3}{4}\right)$

4. Yes.   No.

5. $a = 5$, $c = 6$

6. $x = 2$, $x = -2$, $y = 1$, $y = -1$

   Use calculator to check graph.

Worksheet 8

1. Let f be defined by $f(x) = (x^2 - 1)^3$ for all real numbers x. Without using a calculator, state which of the following is correct.

    (A) $(0, 1)$ relative max

    (B) $(1, 0)$ relative max

    (C) $(-1, 0)$ relative min

    (D) $(0, -1)$ relative min

    (E) $(0, -1)$ relative max, $(-1, 0)$ relative min

2. Given $f(x) - 20 = x^3 - 3x^2 - 9x$ find the absolute maximum value on the closed interval $[0, 6]$.

    (A) 7   (B) 20   (C) 25   (D) 74   (E) 101

3. The figure shows the graph of f', the derivative of the function f. The domain of the function f is $0 \le x \le 10$. For what value(s) is the graph of f concave upwards?

    (A) $0 < x < 4$ (B) $6 < x < 10$

    (C) $2 < x < 6$ (D) $4 < x < 8$

    (E) $0 < x < 8$

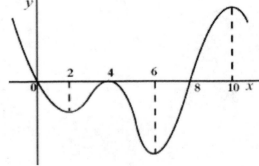

4. The figure shows the graph of f', the derivative of the function f. The domain of the function f is $-10 \leq x \leq 10$. For what value(s) does the function have a possible point(s) of inflection?

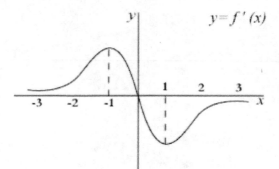

(A) 0   (B) -1 only   (C) 0 only

(D) 1 only   (E) -1 and 1

*5. Given $f(x) = \dfrac{x^4 - 27}{x^2}$. For what values is the graph of f concave down?

(A) $-3 < x < 0$ (B) $0 < x < 3$ (C) $3 < x < \infty$

(D) $-3 < x < 0$ OR $0 < x < 3$ (E) $-\infty < x < -3$ OR $3 < x < \infty$

6. Given $f(x) = 8x^3 - x^4$. For what x values does the graph of f have points of inflection?

(A) 0 only   (B) 4 only   (C) 0, -2   (D) 0, 4   (E) 2, 4

7. The graph $y = \dfrac{x^2 + 3x + 2}{x^2 + 2x - 3}$ has asymptotes at

(A) $x = -3$ and $y = -1$ (B) $x = 1$, $y = -3$, $y = 1$ (C) $y = 1$, $x = 3$, $x = 1$

(D) $y = -1$, $x = -3$, $x = 1$ (E) $y = 1$, $x = -3$, $x = 1$

8. The graph of the **derivative** of f is shown in the figure.

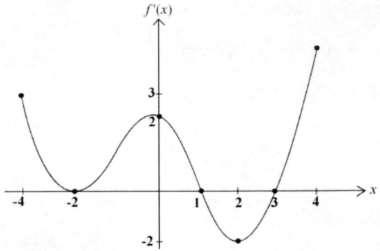

Note: This is the graph of the derivative of f, $y = f'(x)$, for $-4 \le x \le 4$.

a) Suppose that $f(2)=1$. Find an equation of

the line tangent to f at the point (2,1).

b) Where does f have a local minimum? Explain briefly.

c) Where does f have an inflection point? Explain briefly.

d) Where does f achieve its maximum on the interval $[-2,2]$?

9. The graph of f is shown below.

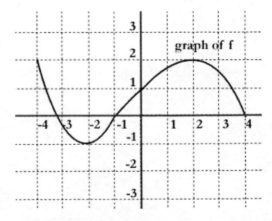

graph of f

Rank the four numbers $f'(0)$, $f(2)$, $f'(2)$, and $f''(2)$ in increasing order.

10. The graph of the **second derivative** of a function f is shown below.

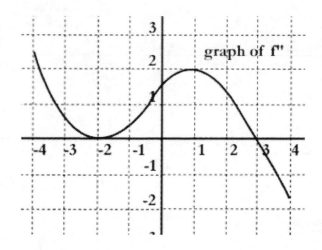

a) Where is the graph of f concave up?

b) Where does f have points of inflection?

c) Suppose $f'(0)=0$. Is f increasing or decreasing at $x=2$? Justify your answer.

11. The slope of a curve at any point (x,y) is given by $\dfrac{dy}{dx}=(x-3)(x-4)^2$.

Determine whether the following statements are *true* or *false*.

a) The curve has a horizontal tangent at the point where $x=3$.

b) The curve has a local minimum at the point where $x=4$.

c) The curve has a local maximum at the point where $x=3$.

d) The curve has an inflection point where $x=4$.

12. Given the following data for a function f.

x	1.1	1.3	1.5	1.7	1.9	2.1
$f(x)$	12	15	21	23	24	25

a) Estimate $f'(1.7)$.

b) Write an equation for the tangent line to the graph of f at $x = 1.7$.

c) Use your answer in b) to predict the value of f at $x = 1.8$.

Answers to Worksheet 8

1. D

2. D

3. B

4. E

5. D

6. D

7. E

8. a) $y + 2x = 5$ b) $x = 3$ c) $x = -2, 0, 2$ d) $x = 1$

9. $f''(2) < f'(2) < f'(0) < f(2)$

10. a) $-4 < x < -2$ OR $-2 < x < 3$

   b) $x = 3$

   c) increasing

11. a) True   b) False   c) False   d) True

12. a) $f'(1.7) \approx 7.5$ b) $y = 7.5(x - 1.7) + 23$ c) 23.75

Worksheet 9

1. f and g are defined as in the graphs below:

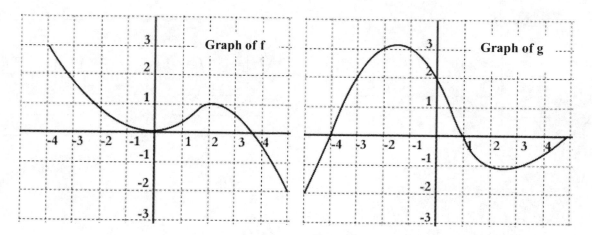

Let h be a function defined by $h(x) = f\big[g(x)\big]$.

a) Evaluate $h(3)$.

b) Is $h'(-2)$ positive or negative? Justify your answer.

c) Estimate $h'(-1)$.

d) Determine the value(s) of x which correspond to critical points of h.

e) Let $h(x) = f(x^2)$. Is h increasing, decreasing, or neither at $x = -1$? Justify

your answer.

2. Given $f(3) = -2$, $f'(3) = 5$ and $g(3) = 3$, $g'(3) = -4$. Find $h'(3)$, if

possible, for each of the following:

a) $h(x) = f(x) \cdot g(x)$

b) $h(x) = \dfrac{f(x)}{g(x)}$

c) $h(x) = f\big[g(x)\big]$

d) $h(x) = \big[f(x)\big]^3$

3. Let $h(x) = f[g(x)]$ and $k(x) = \dfrac{f(x)}{g(x)}$. Fill in the missing entries in the table below.

x	$f(x)$	$f'(x)$	$g(x)$	$g'(x)$	$h(x)$	$h'(x)$	$k(x)$	$k'(x)$
-1	-1	4	1		-1	8	-1	
0	1	0	0	0			2	0
1		-4	1		-1	-8	-1	

4. $y = x^4 + bx^2 + 8x + 1$ has a stationary point of inflection at $x = p$. Find the values of p and b.

5. Find the equation(s) of the tangents to the curve $y = \dfrac{1}{x}$ which pass through the point (-4,2).

6. Find a point <u>not</u> on the graph of $y = x^3 - 1$ from which it is possible to draw <u>EXACTLY</u> <u>TWO</u> tangents to this graph.

<u>Answers to Worksheet 9</u>

1. a) $\dfrac{1}{5}$ b) ≈ -0.5 c) 0.5   d) -4, -3, -1.5, 0, 1, 2.5, 5   e) decreasing

2. a) 23   b) $\dfrac{7}{9}$ c) -20   d) 60

3. $g'(-1) = -2$ $k'(-1) = 2$ $h(0) = 1$ $h'(0) = 0$ $f(1) = -1$ $g'(1) = 2$ $k'(1) = -2$

4. $p = 1$, $b = -6$

5. $y = -x - 2$ and $x + 4y = 4$

6. <u>Any</u> point whose y co-ordinate is -1.

CHAPTER 4

Limits

In the previous chapter it was noted that $y = 3$ is an asymptote to $y = \dfrac{3x-6}{x-1}$.

This means that as $x \to \infty$, $\dfrac{3x-6}{x-1} \to 3$.

We will use the $\to$ symbol to indicate "gets close to".

This could be written $\lim\limits_{x \to \infty} \dfrac{3x-6}{x-1} = 3$.

When we say that the limit of $\dfrac{3x-6}{x-1}$, as x approaches ∞, is 3, we mean that, as x

increases in size, then $\dfrac{3x-6}{x-1}$ gets "closer and closer" to 3.

It does NOT mean that $\dfrac{3x-6}{x-1}$ necessarily equals 3 for any finite value of x.

In earlier study students are told that division by zero is undefined. In a sense this is

true but it is helpful to think that dividing a finite, non-zero value by zero produces a

large value i.e. $\dfrac{1}{0} = +\infty$ or $-\infty$.

For example, $\lim\limits_{x \to 0^+} \dfrac{1}{x} = +\infty$ and $\lim\limits_{x \to 0^-} \dfrac{1}{x} = -\infty$.

The sense in which the non-definedness of division by zero occurs is that it often produces <u>simultaneously</u> a large positive value and a large negative value; clearly a situation which is not defined.

However note that $\lim\limits_{x\to 0}\dfrac{1}{x^2}=+\infty$ whether x is approaching zero from above or below.

In this case it is helpful to think of the limit as defined as $+\infty$.

In cases where the limit is 0 times ∞ or $\dfrac{0}{0}$ or $\dfrac{\infty}{\infty}$ then the numerical value of the limit can be finite or infinite depending on the context. Other examples of non-definedness are

1) $\lim\limits_{x\to 0} 2^{\frac{1}{x}}$ since as $x\to 0^-$, $\lim 2^{\frac{1}{x}}\to 0$

        but as $x\to 0^+$, $\lim 2^{\frac{1}{x}}\to +\infty$.

        i.e. a graph of $y=2^{\frac{1}{x}}$ looks like

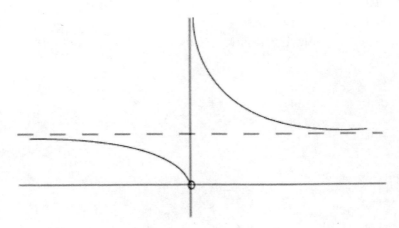

$\therefore$ as $x \to 0$, $2^{\frac{1}{x}}$ approaches 0 and $+\infty$ "simultaneously" which clearly

leads to an undefined value.

2) $\lim\limits_{x \to 0} \dfrac{|x|}{x}$. As $x \to 0^+$, $\dfrac{|x|}{x} \to +1$

but as $x \to 0^-$, $\dfrac{|x|}{x} \to -1$

$\therefore$ as $x \to 0$, $\dfrac{|x|}{x}$ approaches -1 and $+1$ at the same time which again is

not a defined value.

3) $\sin\dfrac{1}{x}$ oscillates wildly as x approaches 0 and hence $\lim\limits_{x \to 0} \sin\dfrac{1}{x}$ is not defined.

4) Similarly, $x\sin x$ oscillates wildly as $x \to +\infty$ and hence $\lim\limits_{x \to \infty}(x\sin x)$ is not defined.

Students often think that saying the limit is undefined is the same as saying the limit is

infinite but this is not necessarily the case. It is helpful to think as follows:

If you go into a shop and enquire about the price of an article that the owner values

highly for personal reasons, he may say, "more than anyone can afford." As a concept

this is intelligible and can be thought of as $+\infty$. If you enquire about the price of

another article and he says, "It costs $6 and $9," then this is clearly an undefined

situation. Undefined means ambiguous or not capable of being explicitly defined. It

does not merely mean infinite.

Examples of Limits

(1) CANCELLING BY A COMMON FACTOR

$\dfrac{10x}{x}$ equals 10 regardless of the value of x.

This is "true" even when x "=" 0. By this we mean that $\lim\limits_{x\to 0}\dfrac{10x}{x}=10$.

This seems reasonable since if $x=0.001$ for example then $\dfrac{10x}{x}=\dfrac{0.01}{0.001}=10$.

It is mathematically correct and legitimate to cancel by a common factor of a

fraction even if the limiting value of that common factor is zero.

For example, $\quad \lim\limits_{x\to 1}\dfrac{x^2-1}{x-1}=\lim\limits_{x\to 1}\dfrac{(x-1)(x+1)}{(x-1)}$

$$=\lim\limits_{x\to 1}(x+1)=2$$

This means that as x gets closer and closer to 1 (from above or below) then

the value of $\dfrac{x^2-1}{x-1}$ gets closer and closer to 2.

(Note for example that when $x=1.01$, $\dfrac{x^2-1}{x-1}=2.01$.)

Note incidentally that it is also legitimate to cancel by $+$ or $-\infty$ provided that

the factors cancelled are the same algebraically.

For example $\lim\limits_{x\to\infty}\dfrac{10x-10}{x-1}=\lim\limits_{x\to\infty}\dfrac{10(x-1)}{x-1}=10$.

(2) <u>ORDER OF SIZE</u>

When we investigate $\lim_{x\to\infty}\dfrac{x^3}{2^x}$ note that as x gets large, x^3 gets large also <u>BUT</u>

2^x gets much bigger still because for example:

When $x=20$, $x^3=8000$ but $2^x=1048576$.

We say that 2^x dominates x^3. In cases like this we think of $\lim_{x\to\infty}\dfrac{x^3}{2^x}$ as $\dfrac{\text{big}}{\text{bigger}}$

which limit approaches 0.

i.e. $\lim_{x\to\infty}\dfrac{x^3}{2^x}=0$.

If we investigate $\lim_{x\to\infty}\dfrac{\log x}{x^2}$, as x gets large, $\log x$ gets large but x^2 gets larger.

$$\therefore \lim_{x\to\infty}\dfrac{\log x}{x^2}=0$$

A rough guide to size is as follows:

logarithmic < polynomial < exponential

Or perhaps

Category	Size
$\dfrac{1}{x}$	smallest
$\log x$	smaller
x	basic
x^2	big
2^x	bigger
x^x	biggest

(3) <u>BOUNDEDNESS</u>

When we consider $\lim\limits_{x\to 0} x\cos\left(\dfrac{1}{x}\right)$, as $x\to 0$ $\dfrac{1}{x}$ gets very large but $\cos\left(\dfrac{1}{x}\right)$ is bounded such that its value lies between -1 and +1 regardless of the value of x.

It follows that $\lim\limits_{x\to 0} x\cos\left(\dfrac{1}{x}\right) = 0$ times 'bounded' which is 0.

i.e. $\lim\limits_{x\to 0} x\cos\left(\dfrac{1}{x}\right) = 0$

(4) Sometimes one can simplify a limit by multiplying the numerator and denominator by a common factor even though, in the limiting case, that common factor equals zero.

For example $\lim\limits_{x\to 1}\left(\dfrac{2-\dfrac{1}{x-1}}{3+\dfrac{1}{x-1}}\right) = \lim\limits_{x\to 1}\dfrac{2x-3}{3x-2}$

$= -1$.

(5) L'Hôpital's Rule

If $\lim\limits_{x \to a} \dfrac{f(x)}{g(x)} = \dfrac{0}{0}$ or $\dfrac{\infty}{\infty}$ then

$\lim\limits_{x \to a} \dfrac{f(x)}{g(x)} = \lim\limits_{x \to a} \dfrac{f'(x)}{g'(x)}$. This is called L'Hôpital's Rule.

For example $\lim\limits_{x \to 3} \dfrac{\sqrt{x+1}-2}{x^2-9} = \lim\limits_{x \to 3} \dfrac{\frac{1}{2\sqrt{x+1}}}{2x} = \dfrac{1}{24}$

Or $\lim\limits_{x \to \infty} \dfrac{x^2+1}{3x^2+3x} = \lim\limits_{x \to \infty} \dfrac{2x}{6x+3} = \lim\limits_{x \to \infty} \dfrac{2}{6} = \dfrac{1}{3}$

Proof of L'Hôpital's Rule in both cases appears at the end of this chapter.

(6) Definition of Derivative

$$f'(x) = \lim\limits_{h \to 0} \dfrac{f(x+h)-f(x)}{h}$$

Note that this means for example that

$\lim\limits_{h \to 0} \dfrac{\sqrt{x+h}-\sqrt{x}}{h}$ is the derivative of $\sqrt{x}$

i.e. $\lim\limits_{h \to 0} \dfrac{\sqrt{x+h}-\sqrt{x}}{h} = \dfrac{1}{2\sqrt{x}}$.

Similarly if $f(x) = x^5$ then $\lim\limits_{h \to 0} \dfrac{(2+h)^5 - 2^5}{h}$ is a definitional

way of writing $f'(2)$.

$f'(x) = 5x^4$ and hence $f'(2) = 5(2^4) = 80$.

It follows that $\lim\limits_{h \to 0} \dfrac{(2+h)^5 - 2^5}{h} = 80$.

Worksheet 1

<u>LIMITS</u>

1. Evaluate the following limits (if they exist)

a) $\lim\limits_{x\to3} \dfrac{x^2+9}{x-3}$

b) $\lim\limits_{x\to3} \dfrac{x^2-9}{x-3}$

c) $\lim\limits_{x\to1} \dfrac{x^2-1}{x^3-x^2}$

d) $\lim\limits_{x\to\infty} x\tan x$

e) $\lim\limits_{x\to\infty} \dfrac{1000x^2}{x^3-5}$

f) $\lim\limits_{x\to\infty} \dfrac{3x-7}{2x+1}$

g) $\lim\limits_{x\to\infty} \dfrac{3+(.2)^x}{(.9)^x+7}$

h) $\lim\limits_{x\to\infty} \dfrac{2^x+x}{3^x}$

i) $\lim\limits_{x\to\infty} \dfrac{\sin x}{x}$

j) $\lim\limits_{x\to\infty} \dfrac{2x^2+3x+5}{3x^2-3x+5}$

k) $\lim\limits_{x\to\infty} \dfrac{2^x-3}{x^4}$

l) $\lim\limits_{x\to64} \dfrac{\sqrt{x}-8}{\sqrt[3]{x}-4}$

m) $\lim\limits_{h\to0} \dfrac{\sqrt{x+h}-\sqrt{x}}{h}$

n) $\lim\limits_{x\to0^+} 2^{\frac{1}{x}}$

o) $\lim\limits_{x\to0^-} 2^{\frac{1}{x}}$

p) $\lim\limits_{x\to\infty} 2^{\frac{1}{x}}$

q) $\lim\limits_{x\to-\infty} 2^{\frac{1}{x}}$

r) $\lim\limits_{x\to2} \dfrac{3-\frac{1}{x-2}}{2+\frac{5}{x-2}}$

s) $\lim\limits_{h\to0} \dfrac{(x+h)^{\frac{1}{4}}-x^{\frac{1}{4}}}{h}$

t) $\lim\limits_{x\to1} 2^{\frac{1}{x}-1}$

u) $\lim\limits_{x\to0} \dfrac{(x+1)^{10}-(x-1)^8}{(x-1)^{10}-(x+1)^8}$

v) $\lim\limits_{x\to0} x\sin\dfrac{1}{x}$

w) $\lim\limits_{h\to0} \left(\dfrac{1}{x+h}-\dfrac{1}{x}\right)\cdot\dfrac{1}{h}$

x) $\lim\limits_{h\to0} \dfrac{(x+h)^2-x^2}{h}$

y) $\lim\limits_{h\to0} \dfrac{(2+h)^2-2^2}{h}$

z) $\lim\limits_{x\to\infty} \dfrac{x^2}{2^x}$

aa) $\lim\limits_{x\to\infty} \dfrac{2^x}{2^{x+1}}$

bb) $\lim\limits_{x\to0} x\cos x$

cc) $\lim\limits_{x\to4} \dfrac{x^2-16}{x^2-7x+12}$

2. $y = \dfrac{ax+b}{x+c}$ looks like the graph shown below.

State possible values for a, b and c.

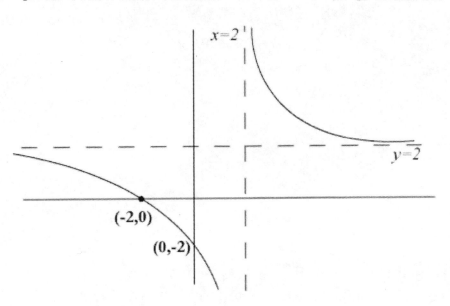

3. Shown is a graph of $y = f(x)$

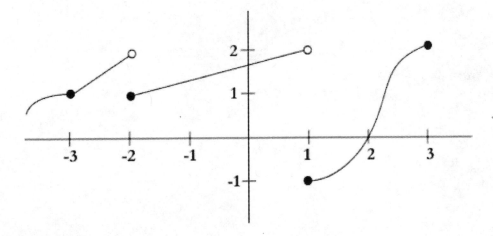

Evaluate the following:

a) $\displaystyle\lim_{x \to 1^-} f(x)$ b) $\displaystyle\lim_{x \to 1^+} f(x)$ c) $\displaystyle\lim_{x \to -2^+} f(x)$ d) $\displaystyle\lim_{x \to -2^-} f(x)$

e) $f(-2)$ f) $\displaystyle\lim_{x \to -2} f(x)$

g) Find the value(s) of a so that $\displaystyle\lim_{x \to a} f(x) = 1$

Answers to Worksheet 1

1. a) undefined b) 6 c) 2 d) undefined

e) 0 f) $\dfrac{3}{2}$ g) $\dfrac{3}{7}$ h) 0

i) 0 j) $\dfrac{2}{3}$ k) ∞ l) 3

m) $\dfrac{1}{2\sqrt{x}}$ n) $+\infty$ o) 0 p) 1

q) 1 r) $-\dfrac{1}{5}$ s) $\dfrac{1}{4}x^{\frac{-3}{4}}$ t) 1

u) -1 v) 0 w) $-\dfrac{1}{x^2}$

x) $2x$ y) 4 z) 0

aa) $\dfrac{1}{2}$ bb) 0 cc) 8

2. $a=2$, $b=4$, $c=-2$

3. a) 2 b) -1 c) 1 d) 2 e) 1 f) undefined g) $a=-3$ or 2.3 (approx.)

Worksheet 2

1. Evaluate the following limits

   a) $\lim\limits_{x\to\infty} \dfrac{x+100}{x^2-9}$

   b) $\lim\limits_{x\to\infty} \dfrac{\sqrt{x}+3}{\sqrt{x}+4}$

   c) $\lim\limits_{x\to\infty} \dfrac{2^x-1}{2^{2x}-1}$

   d) $\lim\limits_{x\to\infty} \dfrac{2^x-10}{x^4}$

   e) $\lim\limits_{h\to 0} \dfrac{(x+h)^3-x^3}{h}$

   f) $\lim\limits_{x\to 0} \dfrac{2^x-1}{4^x-1}$

   g) $\lim\limits_{x\to\infty} \dfrac{3+(.2)^x}{(.9)^x+6}$

   h) $\lim\limits_{x\to\infty} \dfrac{4x^2}{x^2+100x}$

   i) $\lim\limits_{x\to\infty} \dfrac{\sqrt{x}-4}{4-3\sqrt{x}}$

2. Is there a value of c so that $\lim\limits_{x\to 2} \dfrac{2x^2-3cx+x+c}{x^2-4}$ exists? If so, state the value

   of c and the value of the limit.

3. Without resorting to L'Hôpital's Rule, but using a calculator instead,

   investigate whether the following limits exist and if so find their values.

   a) $\lim\limits_{x\to 0} |x|^x$

   b) $\lim\limits_{x\to 0} \dfrac{2x^2}{\sin x+2x^2}$

   c) $\lim\limits_{x\to 0} \dfrac{2x}{\sin x+2x}$

   d) $\lim\limits_{x\to 0} \dfrac{2}{\sin x+2}$

   e) $\lim\limits_{x\to\infty} \sqrt{x^2+4x}-\sqrt{x^2+2x}$

4. The graphs of f and g are shown below.

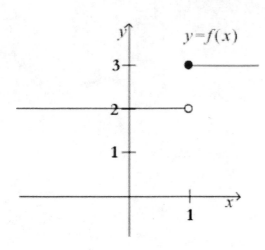

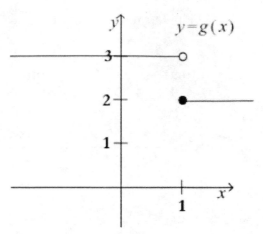

Determine whether the following limits exist and their values if they do exist:

a) $\lim\limits_{x \to 1} f(x)$

b) $\lim\limits_{x \to 1} g(x)$

c) $\lim\limits_{x \to 1} f(x) + g(x)$

d) $\lim\limits_{x \to 1} f(x)g(x)$

e) $\lim\limits_{x \to 1} \dfrac{f(x)}{g(x)}$

f) $\lim\limits_{x \to 1} g(f(x))$

g) $\lim\limits_{x \to 1} \dfrac{g(x)}{f(x)}$

h) $\lim\limits_{x \to 1^+} \dfrac{g(x)}{f(x)}$

i) $\lim\limits_{x \to 1^-} \dfrac{g(x)}{f(x)}$

5. The graph of the derivative of f is shown below.

$$y = f'(x)$$

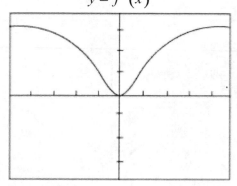

a) Sketch f''.

b) Use the graphs to estimate $\lim_{x \to \infty} f'(x)$ and $\lim_{x \to \infty} f''(x)$.

c) Sketch a graph of f if $f(0)=0$.

6. Let g be the function whose graph is shown below. Use the graph to evaluate each quantity if it exists. If it does not exist explain why.

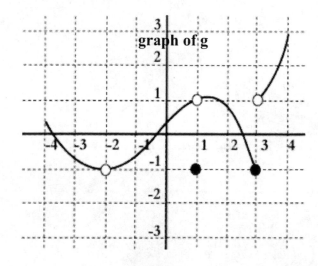

a) $\lim_{x \to -2} g(x)$

b) $\lim_{x \to 1} g(x)$

c) $g(1)$

d) $\lim_{x \to 3^-} g(x)$

e) $\lim_{x \to 3^+} g(x)$

f) $\lim_{x \to 3} g(x)$

7. Let f be the function whose graph if shown below. Use the graph to evaluate each limit, if it exists. If it does not exist explain why.

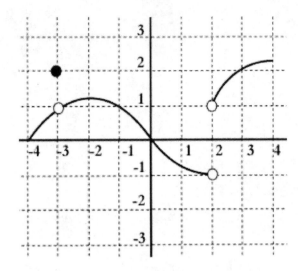

a) $\lim\limits_{x \to -3^-} f(x)$

b) $\lim\limits_{x \to -3^+} f(x)$

c) $\lim\limits_{x \to -3} f(x)$

d) $\lim\limits_{x \to 2^-} f(x)$

e) $\lim\limits_{x \to 2^+} f(x)$

f) $\lim\limits_{x \to 2} f(x)$

8.

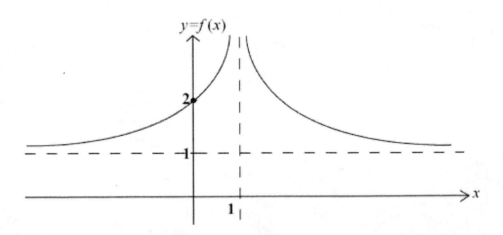

Let $g(x) = f\big(f\big(f(x)\big)\big)$. Find:

a) $\lim\limits_{x \to \infty} g(x)$

b) $\lim\limits_{x \to -\infty} g(x)$

Answers to Worksheet 2

1. a) 0   b) 1   c) 0   d) ∞

    e) $3x^2$ f) $\dfrac{1}{2}$ g) $\dfrac{1}{2}$ h) 4   i) $-\dfrac{1}{3}$

2. $c = 2$, $\text{limit} = \dfrac{3}{4}$

3. a) Yes, 1   b) Yes, 0   c) Yes, $\dfrac{2}{3}$ d) Yes, 1   e) Yes, 1

4. a) No   b) No   c) Yes, 5

    d) Yes, 6   e) No   f) Yes, 2

    g) No   h) Yes, $\dfrac{2}{3}$ i) Yes, $\dfrac{3}{2}$

5. a)   graph of f'' c) graph of f

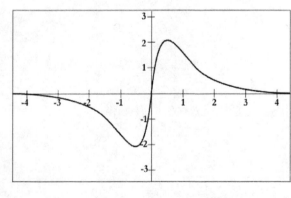 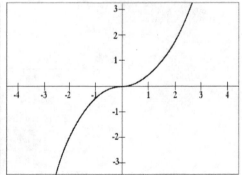

    b) $\lim\limits_{x\to\infty} f'(x) = 3$; $\lim\limits_{x\to\infty} f''(x) = 0$

6. a) -1   b) 1   c) -1

    d) -1   e) 1   f) does not exist

7. a) 1   b) 1   c) 1

    d) -1   e) 1   f) does not exist

8. a) 1   b) 1

Worksheet 3

1. Use the given graph of f to state the value of the limit, if it exists.

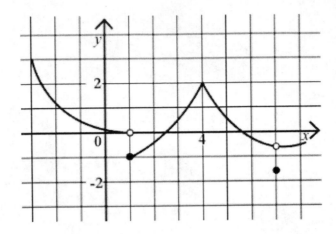

a) $\lim_{x \to -1^-} f(x)$

b) $\lim_{x \to 1^-} f(x)$

c) $\lim_{x \to 1^+} f(x)$

d) $\lim_{x \to 1} f(x)$

e) $\lim_{x \to -3^+} f(x)$

f) $\lim_{x \to 4^-} f(x)$

g) $\lim_{x \to 4^+} f(x)$

h) $\lim_{x \to 4} f(x)$

2.

x	$f(x)$
3.99800	1.15315
3.99900	1.15548
4.00000	1.15782
4.00100	1.16016
4.00200	1.16250

The table above gives values of a differentiable function f. What is the

approximate value of $f'(4)$?

(A) 0.00234   (B) 0.289   (C) 0.427

(D) 2.34   (E) $f'(4)$ cannot be determined from the given

information.

3.

x	-0.3	-0.2	-0.1	0	0.1	0.2	0.3
$f(x)$	2.018	2.008	2.002	2	2.002	2.008	2.018
$g(x)$	1	1	1	2	2	2	2
$h(x)$	1.971	1.987	1.997	undefined	1.997	1.987	1.971

The table above gives the values of three functions, f, g, and h near $x=0$.

Based on the values given, for which of the functions does it appear that the

limit as x approaches zero is 2?

(A) f only   (B) g only   (C) h only (D) f and h only (E) f, g, and h

4.

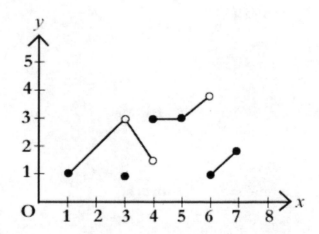

The graph of a function f whose domain is the closed interval $[1,7]$ is shown

above. Which of the following statements about $f(x)$ is true?

(A) $\lim\limits_{x\to 3} f(x)=1$ (B) $\lim\limits_{x\to 4} f(x)=3$ (C) $f(x)$ is continuous at $x=3$

(D) $f(x)$ is continuous at $x=5$ (E) $\lim\limits_{x\to 6} f(x)=f(6)$

5.

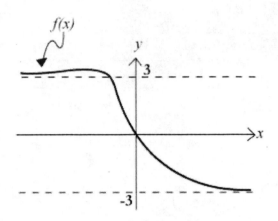

The figure above shows the graph of function $f(x)$ which has horizontal asymptotes $y=3$ and $y=-3$. Which of the following statements is true?

I. $f'(x)<0$ for all $x \geq 0$

II. $\lim_{x \to +\infty} f'(x)=0$

III. $\lim_{x \to -\infty} f'(x)=2$

(A) I only   (B) II only   (C) III only   (D) I and II only   (E) I, II, and III

6. A graph of the function f is shown at the right. Which of the following is true?

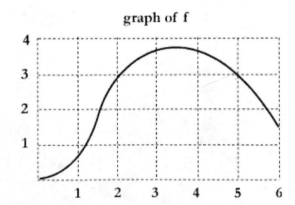

graph of f

(A) $\dfrac{f(5)-f(2)}{5-2}=3$

(B) $\lim_{h \to 0} \dfrac{f(5+h)-f(5)}{h}=2$ (C) $f'(2)<f''(4)$

(D) $f'(1)<f'(3)$ (E) None of these

7. Let P be a point on the graph $y = x^2$. O is the origin (0,0). Let R ($0, r$) be the y intercept of the perpendicular bisector of OP. Find the limit of the value of r as P approaches the origin O.

8. Use your graphing calculator to find a horizontal asymptote to the graph of

   a) $y = \left(x^2 + 4x\right)^{\frac{1}{2}} - \left(x^3 - 6x^2\right)^{\frac{1}{3}}$ b) $y = \left(x^2 + 4x\right)^{\frac{1}{2}} - \left(x^3 - 6x^2 + 12x\right)^{\frac{1}{3}}$

Answers to Worksheet 3

1. a) $\dfrac{1}{2}$ b) 0   c) -1   d) does not exist   e) 3   f) 2   g) 2   h) 2

2. D

3. D

4. D

5. D

6. E

7. limiting value of r is $\dfrac{1}{2}$.

8. a) $y = 4$ b) $y = 4$

Continuity

A function is said to be continuous if it has no 'breaks' in it. In other words a function is continuous over an interval [a,b], if, as we trace over the graph, we do not need to take our pencil off the page. Intuitively, a function f is continuous at a point $(a, f(a))$ if, when x gets close to a, $f(x)$ gets close to $f(a)$.

A function f is continuous at a, if $\lim\limits_{x \to a^+} f(x) = \lim\limits_{x \to a^-} f(x) = f(a)$.

i.e. The graphs below are not continuous at $x = 1$.

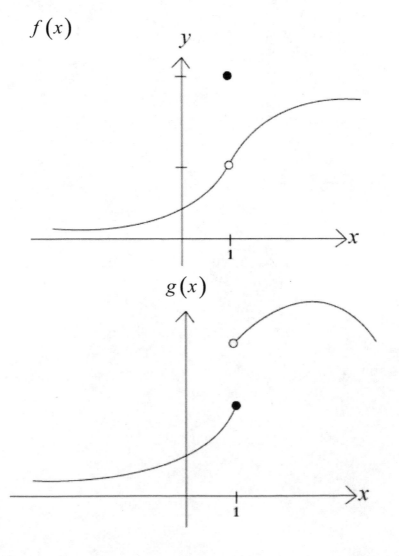

Note that $f(x) = \dfrac{1}{x-1}$ is not continuous $x = 1$ since $\lim\limits_{x \to 1^-} f(x) \neq \lim\limits_{x \to 1^+} f(x)$ and in any case $f(1)$ is not defined.

Since the derivative of a function is itself a function then the same conditions of continuity apply to the derivative.

For example, $f(x) = |x|$ is continuous but $f'(x)$ is not continuous because when $x = 0$ the derivative "jumps" from -1 to +1.

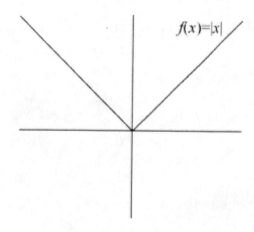

If the derivative of a function is continuous then we say the function is differentiable. Intuitively this means that the function is such that it is possible to draw a unique tangent at every point in the graph of the function.

Note that in the graph of f on page 138 $x = 1$ yields a relative maximum point and an absolute maximum point.

Note that the graph below is not differentiable at $x = -1$ or $x = 3$.

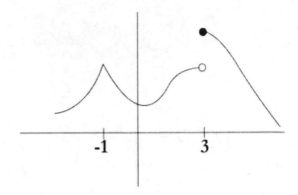

A function which is continuous and differentiable is said to be smooth.

Example

Question: Find the values of a and b so that the function defined by

$$f(x) = 3x^2 + x \ \text{ if } \ x < 2$$

$$f(x) = ax + b \ \text{ if } \ x \geq 2 \qquad \text{is smooth.}$$

Answer: Clearly this function is smooth when $x > 2$ or $x < 2$. Let's look at

$x = 2$. Since the function is continuous then $f(2^-) = f(2^+) = f(2)$.

i.e. $3(2)^2 + 2$ must equal $2a + b$

i.e. $14 = 2a + b$ (1)

Also since the function is differentiable then

$$f'(2^-) = f'(2^+) = f'(2)$$

i.e. $6x + 1 = a$ when $x = 2$

i.e. $13 = a$ (2)

$\therefore \ b = -12, \ a = 13$

Mean-Value Theorem

The Mean-Value Theorem states that if f is a smooth function on an interval $[a,b]$

then there exists a number c between a and b so that

$$f'(c) = \frac{f(b) - f(a)}{b - a}$$

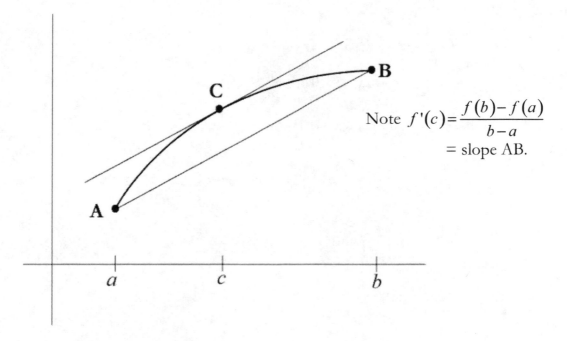

Note $f'(c) = \dfrac{f(b) - f(a)}{b - a}$
$= $ slope AB.

In essence this means that there is a point C (on the curve) between A and B so that

the slope of the tangent at C equals the slope of the straight line AB.

In simplistic language, if you want to go from A to B and you head off (from A) in

the wrong direction then eventually you must turn back towards B to get there.

Example

For $f(x) = x^3 - x^2$ on the interval [-1,3] find a value c between -1 and 3

satisfying the Mean-Value Theorem.

i.e.   we need to solve $3x^2 - 2x = \dfrac{\left(3^3 - 3^2\right) - \left((-1)^3 - (-1)^2\right)}{3 - (-1)}$

i.e. $3x^2 - 2x = 5$

i.e. $3x^2 - 2x - 5 = 0$

$(3x - 5)(x + 1) = 0$

i.e. $x = \dfrac{5}{3}, \ x = -1$

$\therefore$ $f'\left(\dfrac{5}{3}\right)$ and $f'(-1)$ satisfy the MVT.

$\therefore$ $c = \dfrac{5}{3}$.

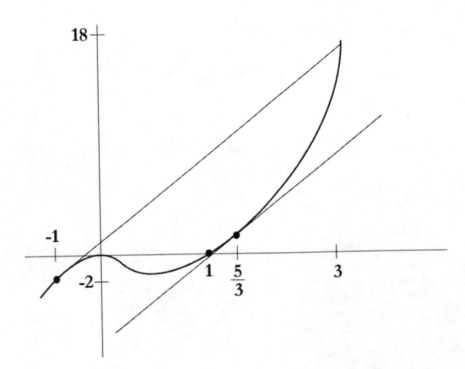

Intermediate Value Theorem

If f is a continuous function between $x = a$ and $x = b$ then for any value M between

$f(a)$ and $f(b)$ there exists at least one value c between a and b so that $f(c) = M$.

Intuitively this means simply that any horizontal line from $y = f(a)$ to $y = f(b)$

must cross the curve at a point whose x co-ordinate is between a and b.

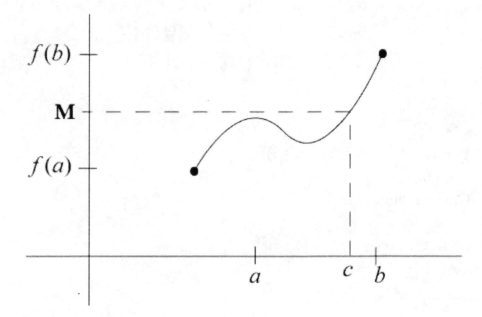

Example

If f is a continuous function on the closed interval [-3,6] and if $f(-3) = -4$

and $f(6) = 10$ then the Intermediate Value Theorem guarantees that

$f(c) = (\text{say})8$ for at least one value , c , between -3 and 6.

Worksheet 4

1. The function g is defined on the interval [-4,4] and its graph is shown below.

   Use the graph of g to answer the following questions.

   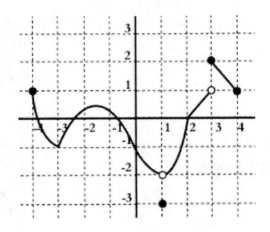

   a) For what values x is g discontinuous?

   b) For what values of x is $g'(x)=0$?

   c) Evaluate $\lim\limits_{x\to 1} g(x)$.

   d) For what values of x does $g'(x)$ not exist? Why?

   e) For what values of x is g' increasing?

   f) Evaluate $\lim\limits_{h\to 0}\dfrac{g(3.5+h)-g(3.5)}{h}$.

2. Let f be the function whose graph is shown to the right. Use the graph to answer the following questions.

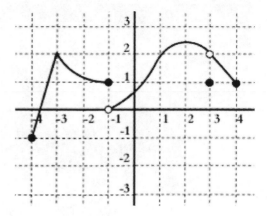

a) For what values of x is f discontinuous?

b) On which intervals is $f'(x) > 0$?

c) Show that f is not continuous at $x = 3$.

d) On which intervals is $f''(x) > 0$?

e) Find $\lim_{x \to -1^+} f(x)$.

3. Let g be the function whose graph is shown to the right. Use the graph to answer the following questions.

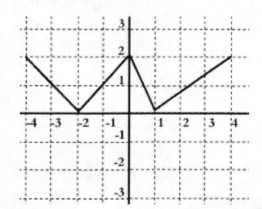

a) Find $g'(-1)$.

b) For what values of x is $g'(x) > 0$?

c) Show that g is continuous at $x = 0$.

d) Find $\lim_{h \to 0} \dfrac{g(2+h) - g(2)}{h}$

4. For what value(s) of b is each of the following functions continuous?

a) $f(x) = \begin{cases} bx^2 - 1, & \text{if } x < 1 \\ x, & \text{if } x \geq 1 \end{cases}$

b) $f(x) = \begin{cases} (x+b)^2, & \text{if } x < 1 \\ (x-2b)^2, & \text{if } x \geq 1 \end{cases}$

c) $f(x) = \begin{cases} \cos(x-1), & \text{if } x \leq 1 \\ \dfrac{b}{x+1}, & \text{if } x > 1 \end{cases}$

5. Let f be the function whose graph is shown to the right. Use the graph to answer the following questions.

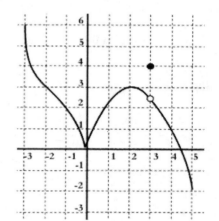

a) Is f continuous on the interval [-3,5]? Justify your answer briefly.

b) Does f have any critical values. Explain briefly.

c) On which intervals is $f'(x) > 0$?

d) On which interval is f' increasing?

e) Identify any inflection points.

f) $\lim\limits_{x \to 0} f(x) =$

g) Is f continuous at $x = 0$? Explain briefly.

h) $\lim\limits_{x \to 2} f'(x) =$

i) $\lim\limits_{h \to 0^+} \dfrac{f(0+h) - f(0)}{h} =$

j) $\lim\limits_{h \to 0^-} \dfrac{f(0+h) - f(0)}{h} =$

k) $\lim\limits_{h \to 0} f'(h) =$

6. Let f be the function whose graph is shown below. Use the graph to answer the following questions.

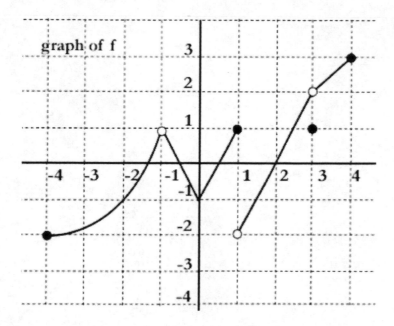

graph of f

a) Is f continuous on the interval $[-4,4]$? Justify your answer briefly.

b) Does f have any critical values? Explain briefly.

c) On which intervals is $f'(x) > 0$?

d) $\lim\limits_{x \to 3} f(x) =$

e) Is f continuous at $x = -1$? Explain briefly.

f) $\lim\limits_{x \to 2} f'(x) =$

g) $\lim\limits_{h \to 0^+} \dfrac{f(0+h) - f(0)}{h} =$

h) $\lim\limits_{h \to 0^-} \dfrac{f(0+h) - f(0)}{h}$

7. Let f be the function whose graph is shown below. Use the graph to answer the following questions.

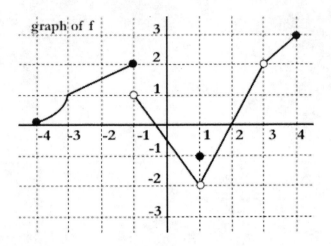

a) Is f continuous on the interval $[-4,4]$? Justify your answer briefly.

b) $\lim\limits_{x \to -1^+} f(x) =$

c) $\lim\limits_{x \to -1^-} f(x) =$

d) On which intervals is $f'(x) < 0$?

e) $\lim\limits_{x \to 0} f(x) =$

f) Is f continuous at $x = 1$? Explain briefly.

8. Let g be the function whose graph is shown below.

   a) State the numbers at which g is discontinuous and explain why.

   b) State the intervals on which g is continuous.

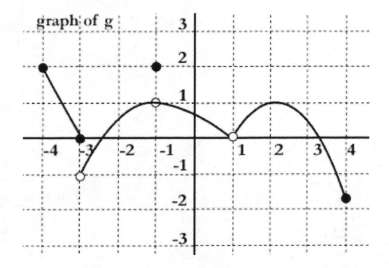

9. Show that if the function g is defined by $g(x) = \begin{cases} \dfrac{x^2 - 9}{5}, & \text{if } x \neq 3 \\ 5, & \text{if } x = 3 \end{cases}$

   then g is not continuous at $x = 3$.

10. Let the function f be defined by $f(x) = \begin{cases} ax^2 + bx, & \text{if } x \leq 3 \\ 3 - 2x, & \text{if } x > 3 \end{cases}$.

   a) For what values of a and b is the function f both continuous and

      differentiable at $x = 3$?

   b) Find values of a and b so that the function f is continuous but not

      differentiable at $x = 3$.

11. If $\lim\limits_{x \to a} f(x) = L$, which of the following statements, if any, MUST be true?

Justify your answer.

   a) f is defined at a.

   b) $f(a) = L$.

   c) f is continuous at a.

   d) f is differentiable at a.

12. If a function f is continuous at a, which of the following statements, if any, MUST be true? Justify your answer.

   a) f is defined at a.

   b) $\lim\limits_{x \to a} f(x)$ exists.

   c) $\lim\limits_{x \to a} f(x) = f(a)$

   d) f is differentiable at a.

13. Determine whether the following statements MUST be true or are at least SOMETIMES false. Justify your answer.

   a) If f is continuous at point A, then it is differentiable at A.

   b) If f is differentiable at point A, then it is continuous at A.

Answers to Worksheet 4

1. a) $x = 1$, $x = 3$ b) $-1\dfrac{1}{2}$ c) -2   d) $x = -3$, $x = 1$, $x = 2$, $x = 3$

   e) [-4,-3] [0,2]   f) -1

2. a) $x = -1$, $x = 3$ b) [-4,-3] (-1,2]

   c) $\lim\limits_{x \to 3^-} f(x) = \lim\limits_{x \to 3^+} f(x) = 2$ but $f(3) = 1$ d) (-3,-1) (-1,1)   e) zero

3. a) 1   b) (-2,0) (1,4)   c) $\lim\limits_{x \to 0^-} g(x) = \lim\limits_{x \to 0^+} g(x) = g(0) = 2$ d) $\dfrac{2}{3}$

4. a) $b = 2$ b) $b = 0$ or $b = 2$ c) $b = 2$

5. a) No. x is discontinuous at $x = 3$ b) $x = 0$, $x = 2$

   c) $0 < x < 2$ d) $-3 < x < -2$ e) $x = -2$

   f) 0   g) Yes   h) 0

   i) +3   j) -4   k) does not exist

6. a) No, the graph is continuous when $x = 1$ and when $x = 3$.

   b) No   c) (-4,-1) (0,1) (1,3) (3,4)   d) 2   e) No   f) 2   g) 2   h) -2

7. a) No. Graph is discontinuous when $x = -1$, $x = 1$, $x = 3$.

   b) 1   c) 2   d) (-1,1)   e) $-\dfrac{1}{2}$ f) No.

8. a) $x = -3$, $x = -1$, $x = 1$ b) [-4,4] except values shown in a)

10. a) $a = -\dfrac{1}{3}$, $b = 0$ b) $b = -3a - 1$ where $a \neq -\dfrac{1}{3}$, $b \neq 0$

11. none of them <u>must</u> be true.

12. I) II) III) must be true   13. a) is sometimes true b) is always true

Worksheet 5

1. Given functions f and g as shown below

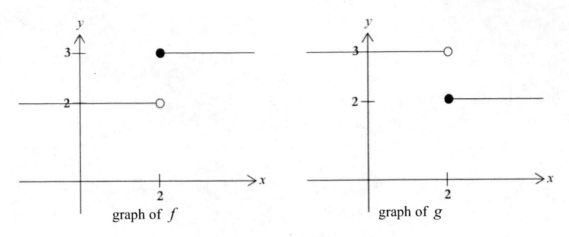

graph of f graph of g

Are the following functions continuous?

a) $f(x)+g(x)$ b) $f(x)g(x)$ c) $f(g(x))$

d) $g(f(x))$ e) $\dfrac{f(x)}{g(x)}$

2. Repeat Question 1 where $f(x)=2$ if $x<3$ and $g(x)=3$ if $x<3$

$f(x)=3$ if $x\ge 3$ and $g(x)=2$ if $x\ge 3$.

3. Let $f(x)=\begin{cases} ax & \text{if } x\le 1, \\ bx^2+x+1 & \text{if } x>1. \end{cases}$

a) Find all choices of a and b such that f is continuous at $x=1$.

b) Draw the graph of f when $a=1$ and $b=-1$.

c) Find the values of a and b such that f is differentiable at $x=1$.

d) Draw the graph of f for the values of a and b found in part c).

4. Hannah takes a trip from Toronto to Montreal. She leaves at 9 AM on Monday and arrives at 2 PM that day. She returns on Tuesday, leaving at 9 AM and arriving back in Toronto, retracing exactly the same route. Show that there is a point on the road through which she passes at the same time both days.

5. Suppose f is continuous and $\lim\limits_{x \to a} f(x) = L$. Find $f(a)$. Justify your answer.

6. Let $f(x) = \begin{cases} 2 & \text{if } x < 0, \\ 3 - x & \text{if } 0 \le x \le 1, \\ x^2 + 1 & \text{if } x > 1. \end{cases}$

   a) Is f continuous at $x = 0$? Justify your answer.

   b) Is f continuous at $x = 1$? Justify your answer.

7. Use the graph of f to estimate the numbers in $[0,8]$ that satisfy the conclusion of the Mean Value Theorem.

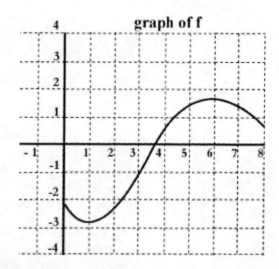

graph of f

8. Use the graph of g to estimate the numbers in $[0,8]$ that satisfy the conclusion

of the Mean Value Theorem.

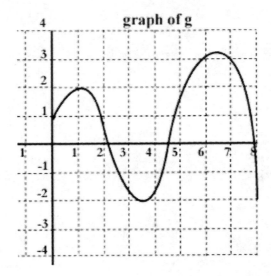

9. Sketch a graph of the function f if $f(x) = \begin{cases} x+2, & \text{if } x \le 1; \\ x^2, & \text{if } x > 1. \end{cases}$

Show that f fails to satisfy the hypothesis of the Mean Value Theorem on the

interval $[-2,2]$ but the conclusion of the theorem is still valid.

10. Determine whether f satisfies the hypotheses of the Mean Value Theorem on the interval $[a,b]$. If it does, find <u>all</u> numbers c in (a,b) such that

$$f'(c) = \frac{f(b) - f(a)}{b - a}$$

   a) $f(x) = 3x^2 + x - 4$ on $[1,5]$

   b) $f(x) = \dfrac{x-1}{x+1}$ on $[0,3]$

   c) $f(x) = \sqrt{x}$ on $[4,9]$

   d) $f(x) = 1 - |x|$ on $[-1,2]$

11. Let $f(x) = x^5 - 3x^2 + 4$. For HOW MANY values of x in the interval from $x = -2$ to $x = 2$ will this function satisfy the conditions of the Mean Value Theorem?

   (A) 0   (B) 1   (C) 2   (D) 3   (E) 4

Answers to Worksheet 5

1. a) Yes   b) Yes   c) Yes   d) Yes   e) No

2. a) Yes   b) Yes   c) No   d) No   e) No

3. a) $a = b + 2$ b) check with calculator   c) $a = 3$, $b = 1$

   d) check with calculator

5. $f(a) = L$ 6. a) No   b) Yes   7. 5.5 or 1.5   8. 1.1, 6.6, 3.4

10. a) $x = 3$ b) $x = 1$ c) $x = \dfrac{25}{4}$

   d) f does not satisfy the differentiability criterion on $[-1,2]$ 11. C

<u>Proof of L'Hôpital's Rule in the Case where $f(a) = g(a) = 0$.</u>

Consider $\lim\limits_{x \to a} \dfrac{f(x)}{g(x)}$ where $f(a) = g(a) = 0$.

Graphically f and g can be represented as shown below.

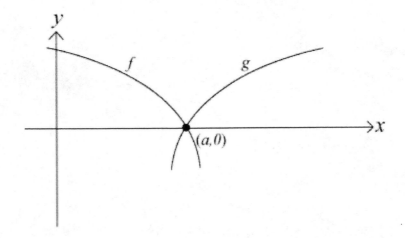

In the limiting case as $x \to a$ we can think of f and g as linear functions.

This means that f can be represented as

$$y - 0 = f'(a)(x - a)$$

i.e $\quad f(x) = f'(a)(x - a)$

and similarly $\quad g(x) = g'(x)(x - a)$

Then $\quad \lim\limits_{x \to a} \dfrac{f(x)}{g(x)} = \lim\limits_{x \to a} \dfrac{f'(a)(x - a)}{g'(a)(x - a)} = \lim\limits_{x \to a} \dfrac{f'(a)}{g'(a)}$

$$= \lim\limits_{x \to a} \dfrac{f'(x)}{g'(x)} \text{ as required.}$$

In the situation where we have

$$\lim_{x \to a} \frac{f(x)}{g(x)} \text{ and } f(a) = g(a) = \infty$$

then the proof is more interesting and complex.

Define $\quad F(x) = \dfrac{1}{f(x)} \text{ and } G(x) = \dfrac{1}{g(x)}$

and let $\quad L = \lim_{x \to a} \dfrac{f(x)}{g(x)}$.

Then $\quad L = \lim_{x \to a} \dfrac{G(x)}{F(x)}$

$$= \lim_{x \to a} \frac{G'(x)}{F'(x)} \text{ (as already shown)}$$

$$= \lim_{x \to a} \frac{(-1)(g(x))^{-2} \cdot g'(x)}{(-1)(f(x))^{-2} \cdot f'(x)}$$

$$= \lim_{x \to a} \frac{[f(x)]^2}{[g(x)]^2} \cdot \frac{g'(x)}{f'(x)}$$

$\therefore \quad L = L^2 \lim_{x \to a} \dfrac{g'(x)}{f'(x)}$

$\therefore \quad \dfrac{1}{L} = \lim_{x \to a} \dfrac{g'(x)}{f'(x)}$

$\therefore \quad L = \lim_{x \to a} \dfrac{f'(x)}{g'(x)}$ ∎

CHAPTER 5

Implicit Differentiation

Implicit differentiation involves a method for finding the slope of a curve whose equation does not represent a function. Note that the use of a graphing calculator is severely limited for this work.

Example

To find the slope of the circle $x^2 + y^2 = 25$ at the point (3,4):

A good way of thinking of this is to recognize that y is related to x and as such the differentiation of y^2, with respect to x, can be considered as an example of the Chain Rule in the sense of $D_x\left[(y)^2\right] = 2y\dfrac{dy}{dx}$.

It follows that if $\qquad x^2 + y^2 = 25$

$\qquad$ then $\qquad 2x + 2y\dfrac{dy}{dx} = 0$.

$\qquad$ At (3,4), $\qquad 2(3) + 2(4)\dfrac{dy}{dx} = 0$

$\qquad$ i.e. $\qquad \dfrac{dy}{dx} = -\dfrac{3}{4}$

Notice that, unlike functions, both the x and y co-ordinates of the point are needed to determine the slope.

When differentiating (say) x^3y^2, with respect to x it should be noted that it is necessary to use the Chain Rule and the Product Rule.

i.e. $D_x\left[x^3y^2\right] = y^2\left(3x^2\right) + x^3 \cdot 2y\dfrac{dy}{dx}$

Example

Find an equation of the tangent to $x^2y^2 + xy^3 = 12$ at the point (1,2).

$$x^2y^2 + xy^3 = 12$$

Differentiate with respect to x.

$$y^2 \cdot 2x + x^2 2y\dfrac{dy}{dx} + y^3 \cdot 1 + x3y^2\dfrac{dy}{dx} = 0$$

At (1,2)

$$8 + 4\dfrac{dy}{dx} + 8 + 12\dfrac{dy}{dx} = 0$$

i.e. $\dfrac{dy}{dx} = -1$

The equation of the tangent at (1,2) is

$$y - 2 = -1(x-1)$$

i.e. $y + x = 3$

Question: Find a relative maximum or relative minimum point on the graph of

$$xy^4 - x^2y^2 = 16.$$

Answer:

$$xy^4 - x^2y^2 = 16 \qquad (1)$$

Differentiate with respect to x:

$$\left(y^4 \bullet 1 + 4xy^3 \frac{dy}{dx}\right) - \left(y^2 \bullet 2x + x^2 2y \frac{dy}{dx}\right) = 0 \quad (2)$$

At a relative max/min point, $\dfrac{dy}{dx} = 0$.

i.e. $\quad y^4 - 2xy^2 = 0$

$$y^2\left(y^2 - 2x\right) = 0$$

i.e. $\quad y = 0 \quad$ or $\quad y^2 = 2x$

Clearly $y = 0$ is not a valid condition for a point to lie on the curve.

When $y^2 = 2x$ we have $x\left(2x\right)^2 - x^2\left(2x\right) = 16$

i.e. $\quad x = 2$.

$\therefore \quad$ A potential relative max/min point occurs when $x = 2$.

When $x = 2$, $y^2 = 4$. $\qquad$ i.e. $y = 2$, or -2.

Note that we can substitute $x = 2$ into $xy^4 - x^2y^2 = 16$ to yield the same result.

$\therefore \quad$ A potential relative maximum/minimum point is

$\left(2, 2\right)$ or $\left(2, -2\right)$.

We need to investigate the sign of $\dfrac{d^2y}{dx^2}$ at $(2,2)$ and $(2,-2)$.

Differentiating (2) with respect to x yields:

$$y^3\frac{dy}{dx}+4y^3\frac{dy}{dx}\cdot1+x\left(\frac{dy}{dx}\cdot12y^2\frac{dy}{dx}+4y^3\frac{d^2y}{dx^2}\right)-\left(2x2y\frac{dy}{dx}+y^2\cdot2+2y\frac{dy}{dx}2x+x^2\left(\frac{dy}{dx}\cdot2\frac{dy}{dx}+2y\frac{d^2y}{dx^2}\right)\right)=0$$

At $(2,2)$ and if $\dfrac{dy}{dx}=0$ we have

$$64\frac{d^2y}{dx^2}-8-16\frac{d^2y}{dx^2}=0 \text{ by substituting in (2').}$$

i.e. $\dfrac{d^2y}{dx^2}=\dfrac{1}{6}$ which is positive

Hence $(2,2)$ is a relative minimum point

Similarly at $(2,-2)$ and if $\dfrac{dy}{dx}=0$, $\dfrac{d^2y}{dx^2}=-\dfrac{1}{6}$.

$\therefore$ $(2,-2)$ is a relative maximum point.

Worksheet 1

1. Find $\dfrac{dy}{dx}$ for each of the following:

   a) $x^2 + y^2 = 4$

   b) $x^2 - xy = 2$

   c) $x^2 + y^2 = 5x - 4y$

   d) $2x^3 - 3y^2 = 7$

2. Find an equation of the tangent to the curve at the given point:

   a) $xy^2 = 16$ at $(4,-2)$

   b) $x^2 - y^2 = 1$ at $(\sqrt{5},2)$

   c) $\sqrt{x} + xy^2 = 18$ at $(4,2)$

   d) $y^3 + 3x^2y + 13 = 0$ at $(2,-1)$

3. Find the value of the constants p and k such that the slope of the tangent to the curve $px^2 + \sqrt{y} + xy = 2k$ is 6 at the point $(1,1)$.

<u>Answers to Worksheet 1</u>

1. a) $\dfrac{-x}{y}$　　　　b) $\dfrac{2x-y}{x}$　　　　c) $\dfrac{5-2x}{2y+4}$　　　　d) $\dfrac{x^2}{y}$

2. a) $4y - x = -12$　　　b) $2y - \sqrt{5}x = -1$　　　c) $64y + 17x = 196$

   d) $5y - 4x = -13$

3. $p = -5,\ k = -\dfrac{3}{2}$

Worksheet 2

1. Find the equation of the tangent to $xy = 1$ at (1,1).

2. Find the equation of the tangent to $x^2 + y^2 = 25$ at (-3,4).

3. Find the equation of the tangent to $\dfrac{x^2}{16} + \dfrac{y^2}{9} = 2$ at (4,3).

4. Find the equation of the tangent to $x^2 y + xy + 2x^2 + 10 = 0$ when $x = 2$.

5. Find the equation of the tangent to $xy^3 (1 - y) = 16$ when $y = 2$.

6. Find the equation of the tangent to $(x + y)^2 = 16x$ at (1,3).

7. Find a relative max or min point for $xy^2 + x^2 y = 2$. State which it is.

8. Find a relative max point on the graph of $9x^2 + 4y^2 - 18x + 18y - 43 = 0$.

9. Find max or min points for $4x^2 - 4xy + y^2 - 12x + 36 = 0$.

10. If $y\dfrac{dy}{dx} = 2x$ and when $x = 0$, $y = 0$ find x when $y = 4$.

11. Graph the following:

   a) $y^2 = x^3$ b) $y^2 = x^3 + 1$ c) $y^2 = x^3 - 1$ d) $y^3 = x^2$

   e) $y^2 = x^2 (12 - x)$ f) $y^2 - x^2 = 4$ g) $y^2 = x^2 (x - 1)$

   h) $y^2 = x(x - 1)^2$ i) $y^2 = x(x - 1)(x - 3)$

12. Find the equation of the tangent to $y^2 = x^2 (12 - x)$ at (3,9). Find where the tangent crosses the curve again.

13. Find the equations of the tangents to $xy^2 - yx^2 = 20$ at the points where $x = 4$.

14. Investigate the max/min points of the graph of $x^{\frac{2}{3}} + y^{\frac{2}{3}} = 1$. Try to graph this relation. State the domain and range and any axes of symmetry.

15. Sketch the following graph showing asymptote(s), intercept(s) and state the domain and range.

$$y^2 = \frac{x+1}{x}$$

Answers to Worksheet 2

1. $x + y = 2$

2. $-3x + 4y = 25$

3. $\dfrac{x}{4} + \dfrac{y}{3} = 2$

4. $6y = 7x - 32$

5. $5y = x + 12$

6. $y = x + 2$

7. $(1, -2)$ Max

8. $(1, 2)$

9. Min at $\left(\dfrac{15}{4}, \dfrac{9}{2}\right)$

10. $+\sqrt{8}$ or $-\sqrt{8}$

11.   a)

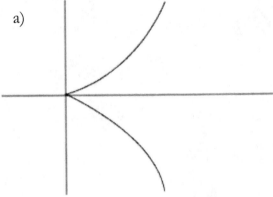

   b)

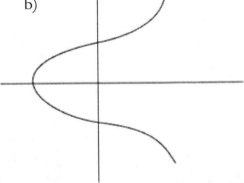

11. c)

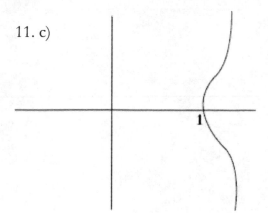

f)

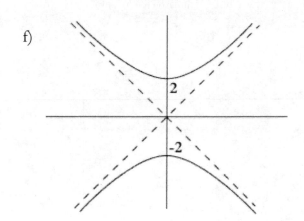

d)

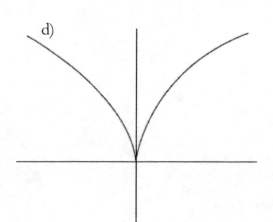

g)

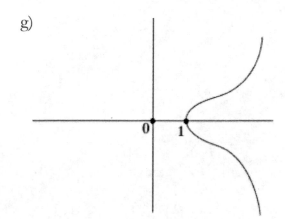

e)

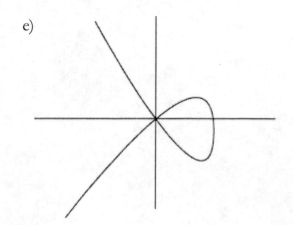

h)

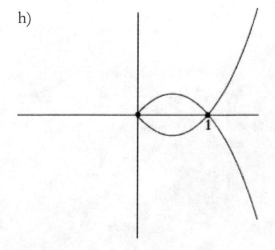

11. i)

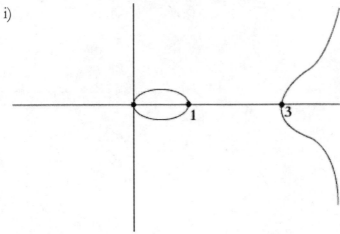

12. $2y = 5x + 3$ $\left(-\dfrac{1}{4}, \dfrac{7}{8}\right)$

13. $8y = 3x - 20$

$8y = 5x + 20$

14. No max or min points

Domain $= \left\{x \mid -1 \le x \le 1\right\}$

Range $= \left\{y \mid -1 \le y \le 1\right\}$

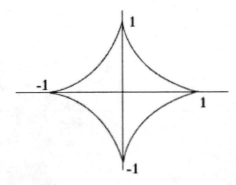

15.

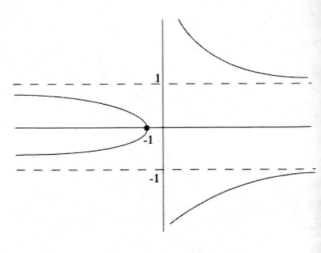

Worksheet 3

IMPLICIT DIFFERENTIATION

1. The tangent to $x^3 + xy + y^2 = 11$ at the point (2,-3) intersects the x-axis at the

   point $(k,0)$. Find the value of k.

2. If $2y\dfrac{dy}{dx} = x$ and $y = 3$ when $x = 2$ find y when $x = 4$.

3. Find the equation of the tangent to $y^2 = x^2(12 - x)$ at the point (8,16). Find

   where the tangent meets the curve again.

4. Find the equation of the tangent to $y^2 = x^2y + 2x$ at (1,2). Does the tangent

   intersect the curve again? If so, where?

5. Sketch $y^2 = x^3 - 4x$ showing the general shape. Label intercept(s) but not

   max/min points or inflection points.

6. If $\dfrac{1}{y^2}\dfrac{dy}{dx} = 2x$ and $y = 1$ when $x = 0$ find the value of y when $x = 2$.

7. State a possible equation for the graph sketched below.

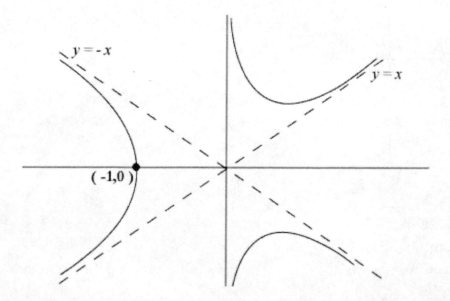

8. For the curve $x^2 - xy + y^2 = 12$, find the point where the tangent is vertical.

9. Let T be an arbitrary point in the first quadrant on the graph of $x^{\frac{2}{3}} + y^{\frac{2}{3}} = 1$.

   Let P and Q be the x and y intercepts of the tangent to graph at point T.

   Prove that the length of PQ is 1 for <u>any</u> point T.

10. Find the equation(s) of the tangents to $x^2 + 4y^2 = 16$ which pass through

   a) the point (2,2)

   b) the point (4,6).

11. Find a relative maximum or relative minimum point on the graph of

   $xy^2 - x^2y = 16$ and state whether the point is a max or a min.

Answers to Worksheet 3

1. $k = \dfrac{10}{3}$ 2. $\pm\sqrt{15}$ 3. $y = 16$ (-4,16) 4. $y = 2x$. Yes at (0,0).

5.

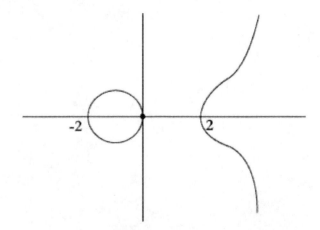

6. $-\dfrac{1}{3}$ 7. $y^2 = \dfrac{x^3 + 1}{x}$ 8. (-4,-2) and (4,2)

10. a) $y = 2$ and $3y + 2x = 10$
    b) $x = 4$ and $3y - 2x = 10$ 11. (2,4) is a relative minimum point.

Worksheet 4

1. A graph of $y^3 + y^2 - 5y - x^2 = -4$ is shown at

   the right.

   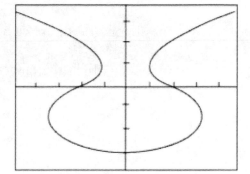

   a) Find $\dfrac{dy}{dx}$.

   b) Write an equation for the line tangent to

   the curve at the point (2,0).

   c) Does the curve have a tangent line at the point (1,1)? Explain your

   reasoning.

2. A curve known as the strophoid is defined

   implicitly by the equation

   $$x^3 + 4x^2 + xy^2 - 4y^2 = 0$$

   and is shown in the diagram. Find the points

   at which the tangent is horizontal. Use your

   graphing calculator.

   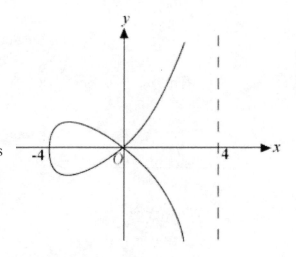

3. Consider the curve $x + xy + 2y^2 = 6$.

   a) Find the equation of the tangent line to the curve at (2,1).

   b) Find all other points on this curve at which the tangent line is parallel

   to the tangent line in part a).

4. The curve $\left(x^2 + y^2\right)^2 = x^2 - y^2$ has the shape of a figure eight. Find the points at which the tangent lines are horizontal.

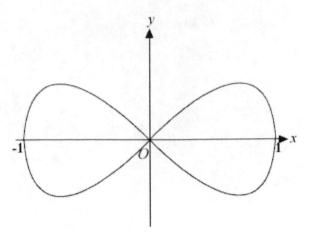

5. If $3x^2 + 2xy + y^2 = 24$ then the value of $\dfrac{dy}{dx}$ when $x = 2$ is

    (A) -2 or 0   (B) 1   (C) 4 or 0   (D) undefined   (E) 3 or -3

6. Find any points on the graph of $4x^2 + xy + 9y^2 = 572$ where the tangent is horizontal. Show clearly where the point(s) you find are max or min points.

Answers to Worksheet 4

1. a) $\dfrac{2x}{3y^2 + 2y - 5}$ b) $y = -\dfrac{4}{5}x + \dfrac{8}{5}$ c) Yes the tangent line is $x = 1$.

2. $(\,-2.47, 1.20\,)$ and $(\,-2.47, -1.20\,)$.

3. a) $3y + x = 5$

    b) $(\,6, -3\,)$

4. $\left(\sqrt{\dfrac{3}{8}}, \sqrt{\dfrac{1}{8}}\right)$ $\left(\sqrt{\dfrac{3}{8}}, -\sqrt{\dfrac{1}{8}}\right)$ $\left(-\sqrt{\dfrac{3}{8}}, \sqrt{\dfrac{1}{8}}\right)$ $\left(-\sqrt{\dfrac{3}{8}}, -\sqrt{\dfrac{1}{8}}\right)$

5. A

6. $(\,1, -8\,)$ $(\,-1, 8\,)$

CHAPTER 6

Maximum / Minimum Problems

Methods for solving practical maximum or minimum problems will be examined by examples.

Example

Question: The material for the square base of a rectangular box with open top costs 27 ¢ per square cm. and for the other faces costs $13\frac{1}{2}$ ¢ per square cm. Find the dimensions of such a box of maximum volume which can be made for $65.61.

Answer:

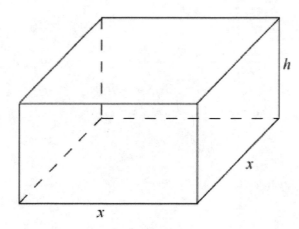

h Let the dimensions of the box be x cms by x cms by h cms as shown.

Cost of making the box is $27x^2 + \left(13\frac{1}{2}\right)4xh$

(base)   (4 faces)

$\therefore \qquad 27x^2 + 54xh = 6561 \qquad (1)$

i.e. $\qquad x^2 + 2xh = 243 \qquad (1)$

———————

We wish to maximize the volume (V) and

$$V = x^2 h \qquad (2)$$

Substituting from (1) into (2) yields:

$$V = x^2 \left(\frac{243 - x^2}{2x} \right) = \frac{1}{2} \left(243x - x^3 \right) \qquad (2)$$

For our purposes, clearly volume and width are positive and a graph of volume against length of the base would look like:

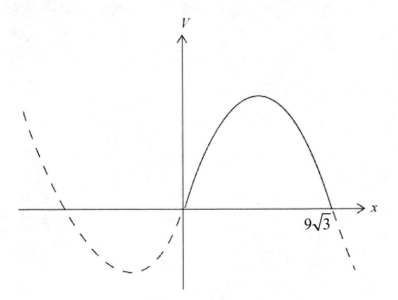

The maximum value occurs when $\frac{dV}{dx} = 0$

i.e.   differentiating (2) with respect to x yields

$$\frac{dV}{dx} = \frac{1}{2} \left(243 - 3x^2 \right)$$

When $\frac{dV}{dx} = 0$, $x = 9$ which clearly yields a maximum volume as seen on the graph. By substituting for x in (1) it follows that $h = 9$ also.

Therefore the box of maximum volume is 9 cms by 9 cms by 9 cms.

Example

Question: Find the point(s) on the graph of $y = x^2$ which is (are) nearest to

$$A\left(0, 1\frac{1}{2}\right).$$

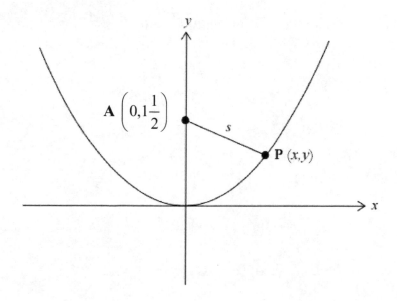

Let P be a point (x, y) on the graph of $y = x^2$. Let $AP = s$.

Then $s = \sqrt{(x-0)^2 + \left(y - 1\frac{1}{2}\right)^2}$

But P lies on $y = x^2$.

$$\therefore \quad s = \sqrt{x^2 + \left(x^2 - 1\frac{1}{2}\right)^2} \quad \quad (1)$$

Note that it would easier to write (1) as $s^2 = x^2 + \left(x^2 - 1\frac{1}{2}\right)^2$ (1) for

differentiating purposes.

We wish to minimize s and hence we need to differentiate (1) with respect to x.

i.e. $\qquad 2s\dfrac{ds}{dx} = 2x + 2\left(x^2 - 1\tfrac{1}{2}\right)2x \qquad\qquad$ (1)'

When s is a relative minimum, $\dfrac{ds}{dx} = 0$.

i.e. $\qquad 0 = 2x + \left(2x^2 - 3\right)2x$

$\qquad\qquad = 2x\left(1 + 2x^2 - 3\right)$

$\qquad\qquad = 2x\left(2x^2 - 2\right)$

$\qquad\qquad = 4x(x-1)(x+1)$

i.e. $\qquad x = 0, \; x = 1, \text{ or } x = -1$.

i.e. $\qquad$ P is $(0,0)$, $(1,1)$, or $(-1,1)$.

Comparing the three distances from A to the three possible positions of P, it is clear that the minimum distance occurs when P is either $(1,1)$ or $(-1,1)$.

Note that, in fact, the distance from $(0,0)$ to A is a relative maximum.

—————————

Sometimes a maximum/minimum question is best answered differentiating more than one equation.

Example

Question: Prove that the rectangle of largest area which can be inscribed in a

circle of fixed radius is a square.

Answer:

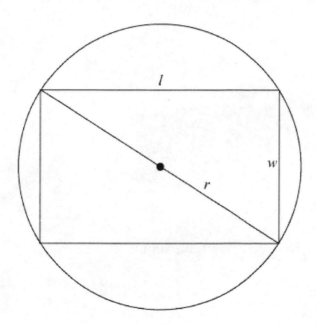

Let the fixed radius of the circle be r and the variable dimensions of

the rectangle be l (length) and w (width).

Let A be the area of the rectangle.

Then $A = lw$ (1)

And $4r^2 = l^2 + w^2$ (2) Pythagoras.

Differentiate <u>both</u> equations with respect to w.

$$\frac{dA}{dw} = w\frac{dl}{dw} + l \qquad (1)'$$

$$0 = 2l\frac{dl}{dw} + 2w \qquad (2)'$$

From (2)', $\dfrac{dl}{dw} = -\dfrac{w}{l}$

Substituting in (1)' yields

$$\frac{dA}{dw} = -\frac{w^2}{l} + l \qquad (1)'$$

When A is maximum, $\dfrac{dA}{dw} = 0$

i.e. $\qquad 0 = -\dfrac{w^2}{l} + l$

and hence $w = l$.

$\therefore \qquad$ The rectangle of maximum area is a square.

Example

Question: A wire of length 60 metres is cut into two pieces. One piece is bent into the shape of an equilateral triangle and the other piece is bent into a square. What are the lengths of each side of the triangle and square so the total area of the triangle and the square is minimized (and maximized?)

Answer:

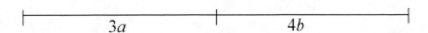

$$3a \qquad\qquad 4b$$

Let each side of the triangle and square be a metres and b metres respectively as shown.

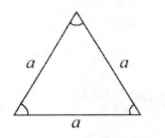 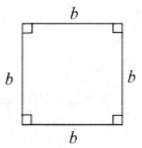

Let A be the total area and

then
$$A = \frac{\sqrt{3}}{4}a^2 + b^2 \qquad\qquad (1)$$

and
$$60 = 3a + 4b \qquad\qquad (2)$$

Differentiate each equation with respect to a.

$$\frac{dA}{da} = \frac{\sqrt{3}}{2}a + 2b\frac{db}{da} \qquad\qquad (1)'$$

$$0 = 3 + 4\frac{db}{da} \qquad\qquad (2)'$$

Substituting for $\dfrac{db}{da}$ from (2)' into (1)' yields

$$\frac{dA}{da} = \frac{\sqrt{3}}{2}a + 2b\left(-\frac{3}{4}\right) \qquad (1)'$$

$$= \frac{\sqrt{3}}{2}\left(a - \sqrt{3}b\right) \qquad (1)'$$

Since A is to be minimized (or maximized)

Let $\dfrac{dA}{da} = 0$ \qquad i.e. \qquad $a = \sqrt{3}b$

and hence substituting in (2) yields $a = 11.3$ (approx.) and $b = 6.524$.

i.e. \qquad Length of side of the triangle is 11.3 and length of the side of

the square is 6.524.

It is not however readily clear whether these dimensions produce a

maximum or minimum total area or possibly only a critical value.

It is clear that a finite length of wire can be a boundary for only a finite

area and hence a maximum area must exist as indeed a minimum area

must exist also.

Substituting for b from (2) into (1) yields

$$A = \frac{\sqrt{3}}{4}a^2 + \left(\frac{60-3a}{4}\right)^2$$

$$\approx 0.9955a^2 - 22.5a + 225$$

Whose graph resembles

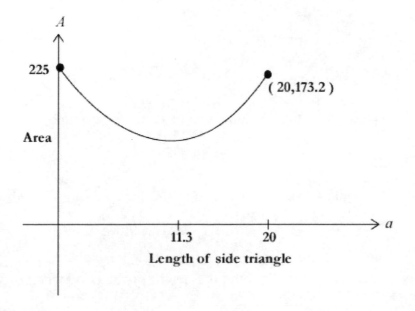

Note that $0 \le a \le 20$ (a bounded domain) and hence from the graph it

is clear that the maximum or minimum total area occurs when $\frac{dA}{da}$ is

not zero, it occurs at an end point of the graph.

To investigate we need to look at the graph of area against the side of

the triangle. The graph clearly illustrates that the <u>minimum</u> total area

occurs when $a = 11.3$ (approx.) as found earlier and the maximum total

area occurs when $a = 0$ i.e. when the piece of wire is bent entirely into

a square (15 by 15) to yield a maximum area of 225 square cms.

Worksheet 1

Max/Min Problems

1. Find the maximum volume of a cylinder whose radius and height add up to 24.

2. The sum of two numbers is 4. Find the maximum value of xy^3 where x and y are the numbers.

3. A rectangular box is to have a capacity of 72 cubic centimetres. If the box is twice as long as it is wide, find the dimensions of the box which require the least material.

4. The volume of a cone is 18π cubic metres. Find the minimum length of the slant edge.

5. The slant edge of a cone is $3\sqrt{3}$. Find the height of the cone when the volume is a maximum.

6. Find the minimum value of $x - \dfrac{8}{3}\sqrt{3x+1}$. Does it have a maximum value?

7. The material for the bottom of a rectangular box with square base and open top costs 3¢ per sq. cm. and for the other faces costs 2¢ per sq. cm. Find the dimensions of such a box of maximum volume which can be made for $5.76.

8. If $xy = 48$ find the minimum value of $x + y^3$ for positive of x and y.

9. Find the dimensions of the cylinder of maximum volume which can be inscribed in a sphere of radius 3 cms.

10. A rectangular sheet of cardboard is 8 cms by 5 cms. Equal squares are cut from each of the corners so that the remainder can be folded into an open topped box. Find the maximum volume of the box.

11. Find the maximum volume of a cylinder which can be inscribed in a cone whose height is 3 cms and whose base radius is 3 cms.

12. The volume of a closed rectangular box with square base is 27 cubic metres. Find the minimum total surface area of the box. Is there a maximum surface area?

13. Find the dimensions of the cone of maximum volume which can be inscribed in a sphere of radius 12 cms.

14. A closed metal box has a square base and top. The square base and top cost $2 per square metre, but the other faces cost $4 per square metre. The minimum cost of such a box having a volume of 4 cubic metres is:

   (A) $2   (B) $8   (C) $16   (D) $48   (E) $64

Answers to Worksheet 1

1. 2048π

2. 27

3. 3 by 4 by 6

4. $3\sqrt{3}$

5. 3

6. $-\dfrac{17}{3}$ No

7. 8 by 8 by 6

8. 32

9. $r = \sqrt{6}$, $h = 2\sqrt{3}$

10. 18 cubic cms

11. 4π

12. 54   No

13. radius of base is $8\sqrt{2}$

   height is 16

14. D

Worksheet 2

MAX/MIN PROBLEMS

1. At 12 noon a ship going due east at 12 knots crosses 10 nautical miles ahead of a second ship going due north at 16 knots.

   a) If s is the number of nautical miles separating the ships, express s in terms of t (the number of hours after 12 noon).

   b) When are the ships closest and what is the least distance between them?

2. Find the minimum distance of a point on the graph $xy^2 = 16$ from the origin.

3. A man can row at 3 m.p.h. and run at 5 m.p.h. He is 5 miles out to sea and wishes to get to a point on the coast 13 miles from where he is now. Where should he land on the coast to get there as soon as possible? Does it matter how far the point on the coast is from the man?

4. Find the dimensions of the rectangle of maximum area in the first quadrant with vertices on the x axis, on the y axis, at the origin and on the parabola $y = 36 - x^2$.

5. Find the maximum volume of a cylinder which can be placed inside a frustrum (lampshade) whose height is 4 cms and whose radii are 1 cms and 3 cms. Is there a minimum volume for the cylinder? If so, what is the radius of that cylinder?

6. A rectangle is to have an area of 32 square cms. Find its dimensions so that the distance from one corner to the mid point of a non-adjacent edge is a minimum.

7. A poster is to contain 50 square cms. of printed matter with margins of 4 cms. each at the top and bottom and 2 cms at each side. Find the overall dimensions if the total area is a minimum. Does the poster have a maximum area?

8. A cylinder has a total external surface area of 54π square cms. Find the maximum volume of the cylinder.

9. Find the shortest distance between $y = x + 10$ and $y = 6\sqrt{x}$.

Answers to Worksheet 2

1. a) $s^2 = (12t)^2 + (10 - 16t)^2$ b) 12:24 p.m.   6 miles

2. $2\sqrt{3}$

3. $3\frac{3}{4}$ miles   No

4. $2\sqrt{3}$ by 24

5. 8π . Yes, $r = 3$.

6. 4 by 8

7. 18 by 9   No

8. 54π cubic cms

9. $4\sqrt{2}$

Worksheet 3

MAX/MIN PROBLEMS

1. *ABCD* is a trapezoid in which *AB* is parallel to *DC*. $AB = BC = AD = 10$ cms. Find *CD* so that the area of the trapezoid is maximized and find the maximum area.

2. A rectangle has constant area. Show that the length of a diagonal is least when the rectangle is a square.

3. A sailing ship is 25 nautical miles due north of a floating barge. If the sailing ship sails south at 4 knots while the barge floats east at 3 knots find the minimum distance between them.

4. A sector of a circle has fixed perimeter. For what central angle θ (in radians) will the area be greatest?

5. The cost of laying cable on land is $2 per metre and the cost of laying cable under water is $3 per metre. In the diagram below the river is 50 metres wide and the distance *AC* is 100 metres. Find the location of P if the cost of laying the cable from A to B is a minimum.

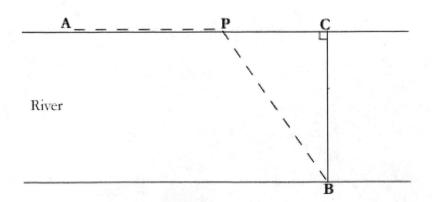

6. A right circular cylindrical can is to have a volume of 90π cubic cms. Find the height h and the radius r such that the cost of the can will be a minimum given that the top and bottom cost 5 ¢ per square cm. and the lateral surface area costs 3 ¢ per square cm.

7. Find the maximum area of a rectangle $MNPQ$ where P and Q are two points on the graph of $y = \dfrac{8}{1+x^2}$ and N and M are the two corresponding points on the x axis.

8. Using a graphing calculator, find the approximate position of the point(s) on the curve $y = x^2 - 4x + 10$ between $(0,10)$ and $(4,10)$

   a) nearest to $(1,6)$.

   b) farthest from $(1,6)$.

Answers to Worksheet 3

1. 20 cms, $75\sqrt{3}$

2. ---

3. 15 miles

4. 2 radians

5. 44.7 miles from C

6. $h = 10$ and $r = 3$

7. 8

8. a) $(1.41, 6.35)$ b) $(4,10)$

Worksheet 4

MAX/MIN PROBLEMS

1. A piece of wire 8 metres long is cut into two pieces. One piece is bent into the shape of a circle and the other into the shape of a square. Find the radius of the circle so that the sum of the two areas is a minimum. Is there a maximum area?

2. Find the dimensions of the rectangle of maximum area in the first quadrant with vertices on the x axis, on the y axis, at the origin and on the parabola $y = 75 - x^2$.

3. A cone has altitude 12 cms. and a base radius of 6 cms. Another cone is inscribed inside the first cone with its vertex at the center of the base of the first cone and its base parallel to the base of the first cone. Find the dimensions of maximum volume.

4. Find the proportions of a right circular cylinder of greatest volume which can be inscribed inside a sphere of radius r.

5. The cost of fuel (per hour) in running a locomotive is proportional to the square of the speed and is $25 per hour for a speed of 25 m.p.h. Other costs amount to $100 per hour regardless of the speed. Find the speed at which the motorist will make the <u>cost per mile</u> a minimum.

6. A motorist is stranded in a desert 5 kms. from a point A, which is the point on a long straight road nearest to him. He wishes to get to a point B, on the road, which is 5 kms. from A. If he can travel at 15 km per hour on the desert and 39 km per hour on the road, find the point at which he must hit the road to get to B in the shortest possible time.

7. A man is 3 miles out to sea from the nearest point A on land on a straight coastline. He can row at 4 m.p.h. and he can jog at k m.p.h. What is his jogging speed if he wishes to reach some point B on the coast as quickly as possible and he therefore lands 4 miles from A? Assume that B is at least 4 miles from A. Note that the distance AB is not relevant.

8. a) Find the point Q on the curve defined by $x^2 - y^2 = 16$ in the interval $4 \le x \le 5$ which is nearest to point P (0,2).

    b) Find the point on the curve in the same interval that is most distant from P.

    c) Verify that PQ is perpendicular to the tangent to the curve at point Q.

Answers to Worksheet 4

1. $\dfrac{4}{4+\pi}$

2. 5 by 50

3. $r = 4$, $h = 4$

4. $h:r = \sqrt{2}:1$

5. 50 m.p.h

6. $\dfrac{25}{12}$

7. 5 m.p.h.

8. a) $\left(\sqrt{17},1\right)$

    b) $(5,-3)$

Worksheet 5

MAX/MIN PROBLEMS

1. An isosceles triangle is circumscribed about a circle of radius 3 cms.

   Find the minimum possible area of the triangle.

2. Find the point on the graph of $y = \sqrt{x}$ which is nearest to $(1,0)$.

3. A variable line through the point $(1,2)$ intersects the x axis at $(a,0)$ and

   intersects the y axis at $(0,b)$. These points are A and B respectively. Find

   the minimum area of triangle AOB if O is the origin and a and b are

   positive.

4. 34 feet of wire are to be divided up into two separate pieces, one of which is

   made into a square and the other into a rectangle which is twice as long at it is

   wide. Find the minimum total area and the maximum total area.

5. What are the dimensions of a rectangle of greatest area which can be laid out

   in an isosceles triangle with base 36 cms. and height 12 cms.

6. The cost of fuel required to operate a boat at a speed of r m.p.h. through the

   water is $0.05r^2$ dollars per hour. If the operator charges \$3 per hour for the

   use of the boat, what is the most economically way, in <u>dollars per mile</u>, to

   travel upstream against a current of 2 m.p.h.

7. A cylindrical vessel with circular base is closed at both ends. If its volume is

   10 cubic centimetres find the base radius when the total external surface is

   least.

8. Find the point(s) on $x^2 - y^2 = 4$ which are closest to $(6,0)$.

Answers to Worksheet 5

1. $27\sqrt{3}$ square cms

2. $\left(\dfrac{1}{2}, \dfrac{1}{\sqrt{2}}\right)$

3. 4

4. 34 square ft and $72\dfrac{1}{4}$ square ft

5. 18 by 6

6. 10 m.p.h. through the water

7. $r = 1.471$ (approx.)

8. $\left(3, \sqrt{5}\right)$ and $\left(3, -\sqrt{5}\right)$

Worksheet 6

1. A man rows 3 miles out to sea from point A on a straight coast. He then wishes to get as quickly as possible to point B on the coast 10 miles from A. He can row at 4 m.p.h. and run at 5 m.p.h. How far from A should he land?

2. An open-topped storage box is to have a square base and vertical faces. If the amount of sheet metal available is fixed, find the most efficient shape to maximize the volume.

3. A teepee is to be made of poles which are 6 metres long. What radius will achieve the teepee of maximum volume?

4. Find the dimensions of the right circular cone of minimum volume which can be circumscribed about a sphere of radius 8 cms.

5. A rectangular sheet of cardboard of length 4 metres and width 2.5 metres has four equal squares cut away from its corners and the resulting sheet is folded to form an open-topped box. Find the maximum volume of the box.

Answers to Worksheet 6
1. 4 miles

2. h by $2h$ by $2h$ where h

4. $h = 32$ and $r = 8\sqrt{2}$

3. $2\sqrt{6}$

5. $2\frac{1}{4}$ cubic m.

CHAPTER 7

Position, Velocity, Acceleration

When we talk of acceleration we think of how quickly the velocity is changing. For example, when a stone is dropped its acceleration (due to gravity) is approximately (9.81 metres/second) per second. This means that every second the velocity increases by 9.81 m/sec. For example, after 1 second, its velocity would be 9.81 m/sec, after two seconds its velocity would be 19.62 m/sec and so on.

If we plotted a velocity time graph it would look like:

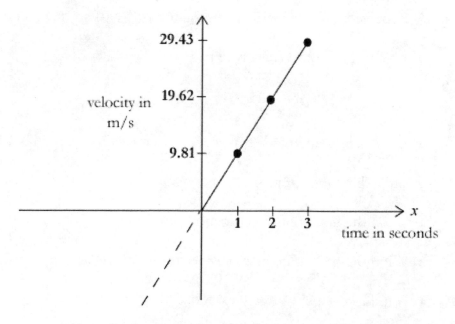

Note that acceleration represents the slope of the velocity/time graph.

This is true even if the velocity time graph is non-linear i.e. acceleration is the derivative of velocity with respect to time.

i.e. $a = \dfrac{dv}{dt}$ where a, v, t refer to acceleration, velocity and time respectively.

Similarly, velocity represents how quickly position is changing and hence

$$v = \frac{dx}{dt} \text{ where } x \text{ denotes position.}$$

Keep in mind that a, v, x are all functions of time and are often represented by

$a(t)$, $v(t)$, $x(t)$. Some texts use $s(t)$ to denote position in place of $x(t)$.

For example, $v(3)$ denotes the velocity after 3 seconds.

Example

Question: A stone is thrown upwards from the top of a cliff so that its position

relative to ground level is given by

$$x(t) = 20t - 5t^2 + 25$$

where t is measured in seconds and x is measured in metres. The

positive direction for position is upwards.

a) Find the height of the cliff above ground level.

b) Find the maximum height above ground level attained by the stone.

c) Find the velocity after 1 second.

d) Find the velocity when the stone hits the ground.

Answer: a) When $t = 0$, the stone is at the top of the cliff.

i.e. $x(0) = 25$. Therefore the height of the cliff is 25 metres.

b) $v(t) = 20 - 10t$

When the maximum height is reached the stone has stopped

moving,   i.e. $v(t) = 0$.

$\therefore$ $20 - 10t = 0$

i.e. $t = 2$.

$\therefore$ The stone stops moving (temporarily) after 2 seconds.

$$x(2) = 20(2) - 5(2)^2 + 25 = 45$$

$\therefore$ The maximum height of the stone is 45 metres above ground

level.

c) The velocity after 1 second $= v(1) = 10$.

d) When the stone hits the ground $x(t) = 0$

i.e. $0 = 20t - 5t^2 + 25$

$= -5(t+1)(t-5)$

i.e. $t = -1$ or 5

In practical situations time is assumed to be positive and clearly we

want $t = 5$.

$v(5) = -30$.

$\therefore$ The stone hits the ground with a speed of 30 m/s. Note that

the minus sign denotes that the stone is travelling downwards.

Speed is understood to be the magnitude of velocity and hence if velocity is -10 m/sc

then the speed is +10 m/s. Speed has no notion of direction.

<u>Example</u>

Question: Consider the motion of a particle such that $x(t) = t^3 - 15t^2 + 8$

for $0 \le t \le 12$. Find the maximum speed.

Answer: $x(t) = t^3 - 15t^2 + 8$

$\therefore \quad v(t) = 3t^2 - 30t$

$\qquad \qquad = 3t(t - 10)$

Graphically this looks like:

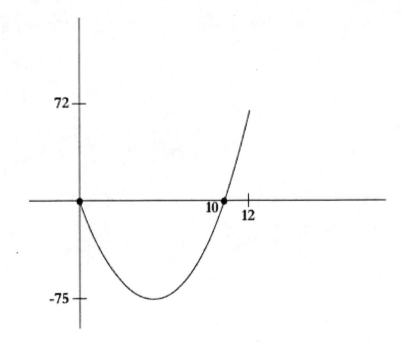

Note that the maximum velocity is 72 m/s but the maximum speed is

75 m/s.

Example

Question: A particle moves in a horizontal line subject to the law

that its position is given by $x(t) = 9t^2 - \dfrac{1}{2}t^4 + 2, \ t \geq 0$.

The positive direction is to the right.

Find :

a) the particle's initial position

b) its velocity after 1 second. Is the particle moving to the right or to

the left at this instant?

c) its position when it first stops. Does it change direction at this time?

d) the maximum or minimum velocity and state which it is.

Answer: a) $x(0) = 2$

$\therefore$ The particle starts at a position of +2 metres.

b) $v(t) = 18t - 2t^3$

$\therefore \quad v(1) = 16$

Since $v(1)$ is positive, the particle is moving to the right (with a

velocity of 16 m/s).

c) When the particle stops, $v(t) = 0$.

    i.e. $0 = 18t - 2t^3$

         $0 = 2t(9 - t^2) = 2t(3 - t)(3 + t)$

    i.e. $t = 0$, 3 or -3.

Clearly we want $t = 3$, and $x(3) = 9(3)^2 - \dfrac{1}{2}(3)^4 + 2 = 42\dfrac{1}{2}$.

$v(3^-)$ is positive and $v(3^+)$ is negative, therefore the particle

changes direction from going to the right to going to the left when

it stops after 3 seconds at a position of $42\dfrac{1}{2}$ m.

d) When velocity is maximum or minimum, $\dfrac{dv}{dt} = 0$.

    i.e.   acceleration $= 0$

        $a(t) = 18 - 6t^2$

            $= 6(3 - t^2)$

When acceleration is zero, $t = +\sqrt{3}$ (remember that time is always

assumed positive) and $v(\sqrt{3}) = 18(\sqrt{3}) - 2(\sqrt{3})^3$

                          $= 12\sqrt{3} = 20.78$ (approx.)

Note that $\dfrac{d^2v}{dt^2} = -12t$ which is negative when $t = \sqrt{3}$ and hence

$v(\sqrt{3}) = 20.78$ m/sec is a <u>maximum</u> velocity.

Example

Question: A particle moves on a number line so that, t seconds after it starts to

move, its velocity in feet per second is $v(t)$,

where $\qquad v(t) = 6t^2 - 12t \qquad t \geq 0.$

The particle starts moving at a position of +4 ft on the number line.

a) Find the initial velocity.

b) Does it move initially to the right or to the left?

c) Find when and where it stops.

d) Find when it moves through the zero position for the first time and

what is its velocity at this time?

e) During what intervals of time is the particle's velocity increasing?

f) Draw a picture (not a graph) of the motion of the particle in the

first 3 seconds.

g) Find the total distance travelled in the first 3 seconds.

h) Find the average velocity during the first 4 seconds.

i) Find the average acceleration during the first 4 seconds.

Answer: a) When $t = 0$, $v(0) = 0$,

therefore the initial velocity is zero.

b) $v(t) = 6t(t - 2)$

When $t = 0^+$, $v(0^+)$ is negative, hence the particle initially moves to

the left.

c) The particle stops when $v = 0$.

i.e. $\quad 0 = 6t(t-2)$

i.e. $\quad t = 0 \text{ or } 2$

$\therefore \quad$ The particle is at rest at the beginning and after 2 seconds.

To find <u>where</u> the particle stops we need to find an expression

for $x(t)$.

Since $\quad v(t) = 6t^2 - 12t$

then $\quad x(t) = 2t^3 - 6t^2 + k$ where k is some constant to be found.

We know the particle is initially at a position of +4, therefore

$$x(0) = 4, \quad \therefore \quad k = 4.$$

$\therefore \quad x(t) = 2t^3 - 6t^2 + 4.$

After 2 seconds, $x(2) = -4$.

Therefore, after starting, the particle stops after 2 seconds at a

position of -4 ft.

d) Since $\quad x(t) = 2t^3 - 6t^2 + 4$

$$= 2(t-1)(t^2 - 2t - 2)$$

When the particle moves through the zero position,

$$x(t) = 0$$

i.e. $\quad t = 1 \text{ or } 1 + \sqrt{3} \text{ or } 1 - \sqrt{3}$ seconds

(the latter two values are found from the Quadratic Formula)

But t is positive, $\therefore$ $t = 1$ or $t = 1 + \sqrt{3}$.

$\therefore$ The <u>first</u> time that the particles moves through the zero position occurs when $t = 1$,

$$v(1) = -6$$

$\therefore$ The particle moves through the zero position for the first time with a velocity of -6 ft/s.

e) The particle's velocity is increasing when acceleration is positive.

$$a(t) = 12t - 12 = 12(t - 1)$$

$\therefore$ velocity is increasing when $t > 1$.

f) We know from b) that the particle moves to the left initially and from c) we know that it stops after 2 seconds at a position of -4. It does not stop anywhere else and after 3 seconds its position is +4 (obtained by substituting $t = 3$ into $x(t) = 2t^3 - 6t^2 + 4$.)

Therefore the motion looks like:

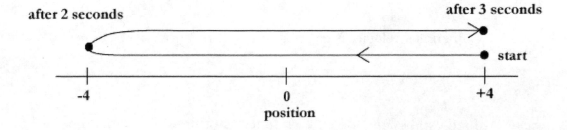

g) Note that the total distance travelled is <u>NOT</u> $x(3)$ which denotes the <u>POSITION</u> of the particle after 3 seconds. From the picture in f) it is clear that the total distance travelled in the first 3 seconds is 16 ft.

h) The average velocity is <u>NOT</u> the averages of the velocities at particular times.

Average velocity means the change in position divided by the time taken (i.e. the change in time).

Average velocity is always measured over a period of time and hence

$$\text{average velocity} = \frac{\Delta x}{\Delta t} \text{ for the time period given.}$$

For our example,

$$\text{average velocity} = \frac{x(4) - x(0)}{4 - 0} = \frac{36 - 4}{4} = 8 \text{ ft/s.}$$

i) Similarly average acceleration $= \dfrac{\Delta v}{\Delta t}$

i.e. average acceleration $= = \dfrac{v(4) - v(0)}{4 - 0} = \dfrac{48 - 0}{4} 12 \text{ ft/s.}$

Example

Sometimes velocity is given in terms of position. For example

$$v(t) = \sqrt{100 - 10x} \quad (1)$$

Question: Find expressions for a, v, x in terms of time, given that the initial velocity is 4 m/sec and the initial position is 2 m.

Answer: $v = \sqrt{100 - 10x}$

$\therefore \quad \dfrac{dv}{dt} = $ derivative of $\sqrt{100 - 10x}$ with respect to <u>TIME</u>

$$= \frac{1}{2}\left(100 - 10x\right)^{-\frac{1}{2}}\left(-10\frac{dx}{dt}\right)$$

$$= \frac{-5}{\sqrt{100 - 10x}}\frac{dx}{dt}$$

$$= \frac{-5}{\sqrt{100 - 10x}}\cdot v \qquad \left(\text{since }\frac{dx}{dt} = v\right)$$

$$= \frac{-5}{\sqrt{100 - 10x}}\sqrt{100 - 10x} \ = -5$$

$\therefore \quad$ acceleration $a(t) = -5$ for all times t.

If $\quad a = -5$

then $\quad v = -5t + k$ where k is some constant

But initially velocity is 4, $\therefore \ v(0) = 4$,

$\therefore \quad k = 4$.

$\therefore \quad v(t) = -5t + 4$

Also, $\quad x(t) = -\dfrac{5}{2}t^2 + 4t + c$ where c is some constant.

But initial position is +2,

Therefore $x(0) = 2$, $\therefore \ c = 2$

$\therefore \quad x(t) = -\dfrac{5}{2}t^2 + 4t + 2$

Speed is defined to be the positive value of velocity for any given time.

i.e. speed = the absolute value of velocity.

For example, on a graphing calculator if we knew that

$$x(t) = t^3 - 2t^2 + t$$

then $v(t) = 3t^2 - 4t + 1$

and speed $= \left| 3t^2 - 4t + 1 \right|$

To find when speed is increasing we could either use our graphing calculator or algebraically we could say that speed is increasing when:

Velocity is both positive and increasing.

or

Velocity is both negative and decreasing.

---

i.e. speed is increasing when velocity and acceleration have the <u>same sign</u>.

Example

Let $x(t) = t^3 - 6t^2 + 9t + 5 \quad (t > 0)$

then $v(t) = 3t^2 - 12t + 9$

$$= 3(t^2 - 4t + 3)$$

$$= 3(t - 1)(t - 3)$$

and $a(t) = 6t - 12$.

To find when speed is increasing we need to look at a signed line diagram.

t	$0 < t < 1$	$1 < t < 2$	$2 < t < 3$	$t > 3$
v	+	-	-	+
a	-	-	+	+

i.e. speed is increasing when $1 < t < 2$ or $t > 3$, since those are the intervals where v and a have the same sign.

Worksheet 1

POSITION, VELOCITY, ACCELERATION PROBLEMS

1. A particle moves such that t seconds after it starts moving it is in a position

    of x metres where $x(t) = t^4 - 6t^3 + 12t^2 - 10t + 3 = (t-1)^3(t-3)$.

a) Find the velocity in metres per second after t seconds.

b) Find the acceleration in metres per second per second after t seconds.

c) Find the initial velocity.

d) Find in which direction the particle starts to move.

e) Find how far to the left it goes.

f) Find when and where the particle changes direction.

2. The position of a locomotive from a fixed point on a straight track at time t

    is x where $x = 3t^4 - 44t^3 + 144t^2$. When was the locomotive moving in

    reverse?

3. A body rises vertically from the earth according to the law $s = 19.6t - 4.9t^2$.

    Show that it has lost one half of its velocity in the first 14.7 metres of its

    motion.

4. A body moves along a horizontal line according to the law $x = t^3 - 9t^2 + 24t$.

    a) When is x increasing?

    b) When is v increasing?

    c) Find the total distance travelled in the first 5 seconds.

5. A particle starts from a point O and moves along a horizontal line so that t seconds later its position relative to O is x inches to the right where $x = 6t - \dfrac{1}{2}t^3$.

a) At what time does it return to O and what is its velocity at that time?

b) What is its position from O when its velocity is zero and what is its acceleration at this time?

c) What is the total distance travelled in the first three seconds?

Answers to Worksheet 1

1. a) $4t^3 - 18t^2 + 24t - 10$

   b) $12t^2 - 36t + 24$

   c) -10

   d) to the left

   e) $-\dfrac{27}{16}$

   f) $s_{\frac{5}{2}} = -\dfrac{27}{16}$

2. $3 < t < 8$

3. ---

4. a) $0 < t < 2$ or $t > 4$

   b) $t > 3$

   c) 28

5. a) $t = 2\sqrt{3}$, $v_{2\sqrt{3}} = -12$

   b) 8, -6

   c) $11\dfrac{1}{2}$

Worksheet 2 – Calculators Required

1. When a ball is thrown straight down from the top of a tall building, with initial velocity -30 ft/sec, the distance in feet above the ground at time t in seconds is given by the formula $s(t) = 300 - 16t^2 - 30t$.

   If the release point is 300 feet above the ground, what is the speed in ft/sec (to the nearest integer) of the ball at the time it hits the ground?

   (A) 142 ft/sec   (B) 145 ft/sec   (C) 149 ft/sec   (D) 153 ft/sec   (E) 157 ft/sec

2. A particle moves along the x-axis so that at time $t > 0$ its position is given by $x(t) = t^4 - 2t^3 - t^2 + 2$. At the instant when the acceleration becomes zero, the velocity of the particle is approximately

   (A) -4.15   (B) -0.594   (C) 0.152   (D) 0.178   (E) 1.985

3. Suppose a particle is moving along a coordinate line and its position at time t is given by $s(t) = \dfrac{9t^2}{t^2 + 2}$. For what value of t in the interval $[1,4]$ is the instantaneous velocity equal to the average velocity?

   (A) 2.00   (B) 2.11   (C) 2.22   (D) 2.33   (E) 2.44

4. A particle is moving along the x-axis so that at time t its velocity is given by

$v(t) = 3t^2 - 2t$. At the instant when $t = 0$, the particle is at the point $x = 2$.

The position of the particle at $t = 3$ is:

(A) 12   (B) 16   (C) 20   (D) 24   (E) none of these

5. In the graph shown below of position, state at which of the points A, B, C, D, or E the speed is the greatest.

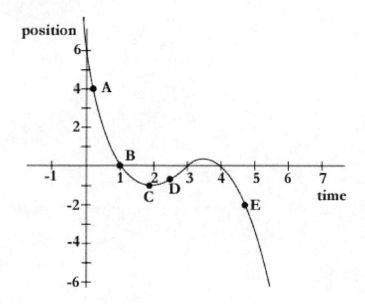

6. A particle moves along the x-axis so that its position $0 \le t \le 5$ is given by

$x(t) = t^4 - 15t^3 + 44t^2 - 36t + 2$. For which value of t is the **speed** the greatest?

(A) $t = 1$ (B) $t = 2$ (C) $t = 3$ (D) $t = 4$ (E) $t = 5$

7. A particle moves along the x-axis in such a way that its position at time t is given by the formula $x(t) = 3t^5 - 25t^3 + 60t$, $t \geq 0$. For what values of t is the particle moving to the left?

(A) $0 < t < 1$ (B) $0 < t < 1$ or $1 < t < 2$ (C) $0 \leq t < 2$

(D) $1 < t < 2$ (E) $t > 2$ or $0 \leq t < 1$

8. The graph shows the <u>position</u> of a toy car after t seconds.

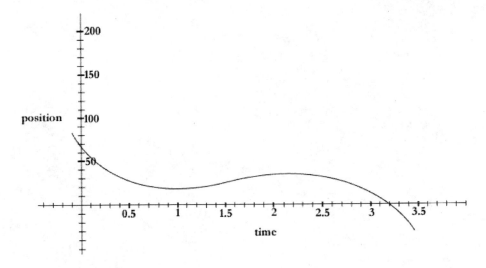

a) At which time in the interval $[0,3]$ is the acceleration equal to 0?

(A) 0   (B) 0.5   (C) 1.6   (D) 2.5   (E) 3.2

b) At which time in the interval $[0,3]$ could the average velocity of the car over that time period be equal to its instantaneous velocity?

(A) 0.8   (B) 1.2   (C) 1.5   (D) 1.8   (E) 2

9. If $v = \dfrac{2}{\sqrt{t}} - \dfrac{t}{4}$ and the initial position is $s = 30$, find the position at the time when the particle is stationary.

    (A) 6   (B) 24   (C) 36   (D) 40   (E) 12

10. Find when speed is increasing if $x(t) = 2t^3 - 21t^2 + 60t + 8$.

11. Find when speed is increasing if $x(t) = 2^t - t^2$, $t \geq 0$. You will need your graphing calculator.

Answers to Worksheet 2

1. A

2. A

3. C

4. C

5. A

6. E

7. D

8. a) C   b) A

9. C

10. $2 < t < 3\dfrac{1}{2}$ or $t > 5$.

11. $0.485 < t < 2.0575$ or $t > 3.2124$.

Worksheet 3

POSITION, VELOCITY, ACCELERATION

*1. A particle moves such that t seconds after it starts to move its velocity is given by the formula $v = 2t^3 - 14t^2 + 22t - 10$ where v is measured in feet per second. Initially the particle is at a position of 3 feet.

a) Find its position after 2 seconds.

b) Find the times during which it has increasing velocity.

c) Find the times during which it is moving to the right.

d) Find the times and positions when the particle is at rest.

e) Find the times and positions when the particle changes direction.

f) Find the total distance travelled in the first 5 seconds.

2. The position s of a particle on a number line at the end of t seconds is given by $s = t^4 - 4t^3 + 8t - 3$ where t is positive.

a) Find the initial velocity.

b) Find the maximum or minimum velocity and state which it is.

c) Find when the particle stops. Does it change direction at these times?

3. Find the minimum velocity of a body which moves such that its position s feet after t seconds is given by $s = t^4 - 8t^3 + 18t^2 + 2t + 7$.

4. The position of a moving particle from a given fixed point O is given by $s = 2t^3 - 15t^2 + 24t + 8$ where t is in seconds and s is in metres. Find the acceleration at the time when the position of the particle relative to O is a minimum.

5. The velocity of a particle is given by the equation $10v^2 = 64s + 1000$ where s is the position in feet from the starting point. Find the acceleration if v is measured in feet per second.

6. The acceleration of a rocket after t seconds from lift-off is given by the formula $a = 15\sqrt{t+100} - 10$ metres per second per second. If the initial position and velocity are both zero

   a) Find the position and velocity in terms of t.

   b) Find the velocity after 2 seconds.

Answers to Worksheet 3

1. a) $-\dfrac{7}{3}$

   b) $0 < t < 1$ or $t > 3\dfrac{2}{3}$

   c) $t > 5$

   d) $s_1 = -\dfrac{1}{6}$, $s_5 = -42.8$

   e) $s_5 = -42.8$

   f) 45.83

2. a) 8

   b) minimum velocity occurs

   when $t = 2$, $v_1 = -8$

2. c) $t = 1$ and $t = 1 + \sqrt{3}$

   It changes directions both times.

3. 2 -note its minimum velocity is its

   initial velocity and again when t=3

4. +18

5. $3\dfrac{1}{5}$

6. a) $v = 10(t+100)^{\frac{3}{2}} - 10t - 10000$

   $s = 4(t+100)^{\frac{5}{2}} - 5t^2 - 10000t - 400000$

   b) 281.5

Worksheet 4 (No Calculators)

1. Suppose that $s(t) = t^2 - t + 4$ is the position function of the motion of a particle moving in a straight line.

   a) Explain why the function s satisfies the hypothesis of the Mean Value Theorem.

   b) Find the value of t in $[0,3]$ where the instantaneous velocity is equal to the average velocity. [Note how this question mirrors the statement of the Mean Value Theorem]

2. The graph below shows the distance $s(t)$ from zero of a particle moving on a number line, in terms of time. At what time does the acceleration first become negative?

   (A) 1   (B) 3   (C) 5   (D) 7   (E) 9

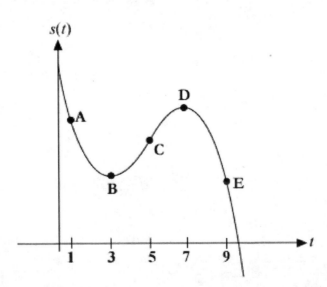

3. The figure shows the graph of the velocity of a moving object as a function of time. At which of the marked points is the object farthest to the right?

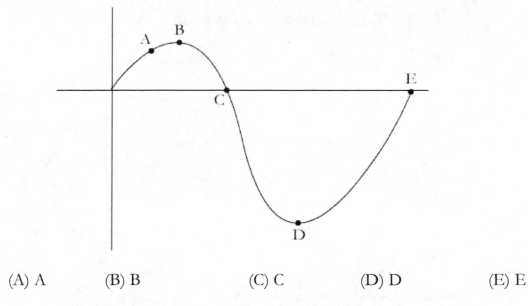

(A) A   (B) B   (C) C   (D) D   (E) E

4. A particle moves along the x-axis in such a way that its **acceleration** in metres per second per second at time t seconds is given by the formula: $a(t) = 2t - 1$. If the initial velocity is -6 m/s, at which of the following (in seconds) is the particle stationary?

(A) $\dfrac{1}{2}$ only   (B) 1 and 3   (C) 3 only   (D) 2 and 3   (E) 4 only

5. The motion of a swimmer playing in the water swimming straight out (or in) from a point on the shore is described by the formula:

$s(t) = 5(t-4)^2 (t-1) + 70$, valid only for $0 \leq t \leq 4$. The position s is in metres, t is in minutes. The velocity of the swimmer is negative when

(A) $0 < t < 2$ (B) $1 < t < 4$ (C) $0 < t < 3$ (D) $2 < t < 4$ (E) $2 < t < 3$

6. At $t = 2$, which of the following statements is true for the motion described by $s = -t^3 + t + 12$:

        I.   The particle is moving toward the zero position.

        II.   The speed of the particle is increasing.

        III.   The velocity of the particle is decreasing.

(A) I only   (B) II and III only   (C) I, II, III

(D) I, II only   (E) I, III only

Answers to Worksheet 4

1. s is continuous and differentiable. $t = 1.5$

2. C

3. C

4. C

5. D

6. C

Worksheet 5

1. A particle moves in a horizontal line according to the law $x(t) = \dfrac{1}{2}t + \dfrac{1}{1+t^2}$

   where $x(t)$ denotes the position after time t seconds.

   a) Find the initial velocity.

   b) Does the particle stop? If so, when?

   c) Find the average velocity between time $t = 1$ sec and $t = 2$ sec.

2. A particle moves along the x-axis in such a way that its position in metres at

   time t seconds is given by the formula $x(t) = \dfrac{1-t}{1+t}$. What is the acceleration

   of the particle at $t = 0$?

   (A) $-\dfrac{3}{5}$ (B) -4   (C) 4   (D) 2   (E) -2

3. Let $s(t) = t^3$. For which value(s) of t in the interval $[0,3]$ does the average

   velocity equal the instantaneous velocity?

   (A) 0   (B) 1   (C) $\sqrt{3}$ (D) 2   (E) 3

4. The velocity of a particle is given by $v = \sqrt{36 - x}$ where x represents the

   position of the particle after time t seconds. The initial velocity of the particle

   is 5 units per second.

   a) Show that the acceleration of the particle is constant.

   b) Express velocity and position as functions of time.

   c) Find the initial position of the particle.

5. The acceleration of a particle is given by $a(t) = 15\sqrt{t+1}$. The initial position of the particle is 5 units and the initial velocity is 6 units per second. Find the position and velocity as functions of time.

6. A particle moves so that its position $x(t)$ after t seconds is given by

   $x(t) = (t-2)^2(t-4)^2$ where $0 \leq t \leq 4$.

   a) When and where does the particle stop?

   b) Does it change direction at the times it stops?

   c) Is there a maximum velocity? If so, at what time does it occur?

   d) Is there a minimum acceleration? If so, what is that minimum acceleration?

7. A particle moves along a horizontal line so that its velocity $v(t)$ at time t seconds is given by $v(t) = 3(t-2)(t-6)$ where $0 \leq t \leq 8$. At time $t = 2$ seconds, the position of the particle is $x(2) = 30$.

   a) Find the minimum acceleration.

   b) Find the total distance travelled by the particle.

   c) The average velocity of the particle over the interval $0 \leq t \leq 5$.

8. The velocity $v(x)$ of particle is given by $v(x) = \sqrt{144 - 4x}$ and the initial velocity of the particle is 12 units per second.

   a) Find the acceleration and velocity as functions of time.

   b) Find the initial position of the particle.

9. Two particles have positions at time t seconds given by $x_1(t) = 4t - t^2$ and

$x_2(t) = 5t^2 - t^3$.

a) Find the velocities $v_1(t)$ and $v_2(t)$ at the instant when their accelerations

are equal.

b) Find the maximum distance between the particles in the interval $[0,4]$.

10. A particle moves so that its position is given by $x(t) = -t^4 + 8t^3 - 18t^2 + 5t + 6$.

Find the times when the velocity is increasing.

Answers to Worksheet 5

1. a) $\dfrac{1}{2}$ b) Yes, it stops after .296 and also after 1 second   c) $\dfrac{1}{5}$ m/s.

2. C

3. C

4. a) $a = -\dfrac{1}{2}$ b) $v = -\dfrac{1}{2}t + 5$, $x = -\dfrac{1}{4}t^2 + 5t + 11$ c) 11

5. $v = 10(t+1)^{\frac{3}{2}} - 4$, $x = 4(t+1)^{\frac{5}{2}} - 4t + 1$

6. a) $t = 2$, $t = 3$, $t = 4$ b) Yes all three times.   c) $t = 2.42$.

   d) Yes when $t = 3$. Minimum acceleration is -4.

7. a) -24   b) 96   c) 1

8. a) $a = -2$, $v = -2t + 12$ b) 0

9. a) $v_1 = 0$, $v_2 = 8$ b) 16.71 units at $t = 3.633$ seconds

10. $1 < t < 3$

CHAPTER 8

Related Rates

When we talk of acceleration we mean the rate at which velocity is changing. If the acceleration is large then, for example, we might say "the velocity is changing quickly". This refers to "how the velocity changes as <u>time</u> changes" and is written $\frac{dv}{dt}$.

Similarly we may talk of how quickly the angle of the sun changes during the day. This could be written as $\frac{d\theta}{dt}$ where θ is the angle of elevation of the sun.

When a ladder slides down a wall as shown in the diagram:

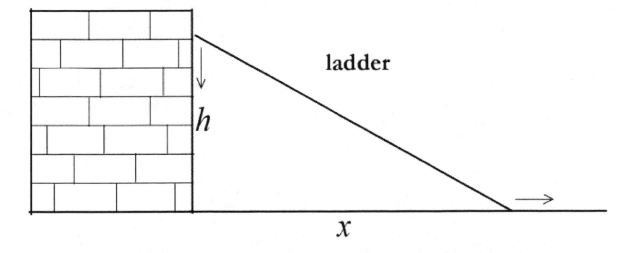

then $\frac{dh}{dt}$ and $\frac{dx}{dt}$ represent how quickly the height is changing and how quickly the distance of the base of the ladder from the wall is changing.

Each of these is an example of what we call related rates. Related rate problems all have the common characteristic that variables are differentiated with respect to <u>TIME</u>.

The expression "how quickly", "at what rate", "how fast is the … changing", "rates of change" all suggest rates.

Remember that the rate at which a variable is increasing also can well be a variable. For example, when a child blows up a balloon, presumably puffing at a constant rate, the rate at which the radius of the balloon increases is much greater when the child first starts puffing than later on at the point when the balloon is about to burst.

The basic approach for solving related rate problems is to write a general equation relating the variables and then differentiate with respect to <u>time</u>. Only then are specific values for the variables substituted.

Example

Question: The volume of a cube is changing at the constant rate of

75 cubic cm/min.

a) Find the rate of change of an edge of the cube when the length of

the edge is 5 cm.

b) Find also the rate of change of the surface area when the surface

area is 24 square cm.

Answer: a) Let V and A represent the volume and surface area of the cube

respectively.

We know that $\dfrac{dv}{dt} = 75$ since this is a <u>constant</u> rate. We are trying

to find $\dfrac{dx}{dt}$ at the instant when $x = 5$.

$$V = x^3 \qquad (1)$$

Differentiate with respect to time

$$\dfrac{dV}{dt} = 3x^2 \dfrac{dx}{dt} \qquad (1)'$$

At the particular instant when $x = 5$ we have

$$75 = 3(5)^2 \dfrac{dx}{dt}$$

i.e. $\qquad \dfrac{dx}{dt} = 1$

This means that, at the instant when the length of the edge of the

cube is 5 cm, the edge is increasing at a rate of 1 cm/sec.

b) $\qquad A = 6x^2 \qquad (2)$

$\therefore \qquad \dfrac{dA}{dt} = 12x \dfrac{dx}{dt} \qquad (2)'$

We need to determine the length of the edge of the cube at the

instant when the surface area = 24 square cm.

i.e. $\quad 24 = 6x^2$

$\qquad 2 = x$

It is <u>NOT</u> valid to substitute in 1 for $\dfrac{dx}{dt}$ in (2)' from part a) because

$\dfrac{dx}{dt}$ is only equal to 1 at the instant when $x = 5$.

We need to determine $\dfrac{dx}{dt}$ at the instant when $x = 2$ from (1)'.

i.e. $75 = 3\left(2^2\right)\dfrac{dx}{dt}$

i.e. $\dfrac{25}{4} = \dfrac{dx}{dt}$

Substituting for x and $\dfrac{dx}{dt}$ in (2)' we have:

$$\dfrac{dA}{dt} = 12(2)\left(\dfrac{25}{4}\right) = 150 \text{ square cm/minute.}$$

Note, as already mentioned, that it is only valid to substitute in (given) special values of variables <u>AFTER</u> differentiating. This is analogous to finding the equation of the tangent to $y = x^2$ at $(2,4)$. To find the slope one may only substitute in 2 for x after differentiating.

Example

Question: The edges of a right-angled triangle are changing but the perimeter is fixed at 40 cm. When the hypotenuse is changing at a rate of 7 cm/min and the edges are 8 cm, 15 cm, 17 cm, find the rates of change of the other two edges at this instant.

Answer: Let the edges be x, y, z cm then

$$x^2 + y^2 = z^2 \qquad (1)$$

and $\quad x + y + z = 40 \qquad (2)$

It is easier to differentiate both equations (1) and (2) with respect to TIME.

i.e. $\quad 2x\dfrac{dx}{dt} + 2y\dfrac{dy}{dt} = 2z\dfrac{dz}{dt} \qquad (1)'$

and $\quad \dfrac{dx}{dt} + \dfrac{dy}{dt} + \dfrac{dz}{dt} = 0 \qquad (2)'$

Substituting x, y, z and $\dfrac{dz}{dt}$ at the special instant we have:

$$16\dfrac{dx}{dt} + 30\dfrac{dy}{dt} = 34(7) \qquad (1)'$$

and $\quad \dfrac{dx}{dt} + \dfrac{dy}{dt} + 7 = 0 \qquad (2)'$

Solving (1)' and (2)' yields $\dfrac{dx}{dt} = -32$ cm/min and $\dfrac{dy}{dt} = 25$ cm/min.

This means that the shorter edge of the right angled triangle is decreasing at 32 cm/min and the longer edge is increasing at 25 cm/min.

Example

Question: A basin has the shape of an inverted cone with altitude 100 cm and

radius at the top of 50 cm. Water is poured into the basin at the

constant rate of 40 cubic cm/minute. At the instant when the volume

of water in the basin is 486π cubic centimetres, find the rate at which

the level of water is rising. $(V = \frac{1}{3}\pi r^2 h)$

Answer:

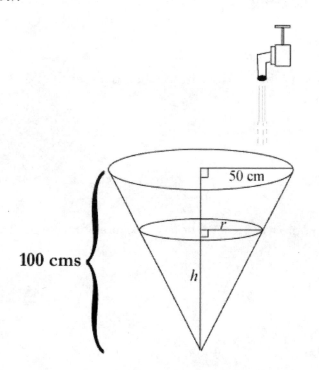

Let the radius of the
top of the water be r
cm and let the water's
depth be h cm as
shown.

Let V be the volume of the water.

We know $\dfrac{dV}{dt} = 40$. We wish to find $\dfrac{dh}{dt}$ at the instant when

$V = 486\pi$ cubic cm.

$$V = \frac{1}{3}\pi r^2 h \qquad (1)$$

$$\frac{r}{h} = \frac{50}{100} \qquad (2) \qquad \text{similar triangles}$$

i.e. $\qquad r = \frac{h}{2} \qquad (2).$

In the previous example we differentiated <u>BOTH</u> equations with respect to time. Here, it is recommended (but not required) that we substitute r, in terms of h, from (2) into (1).

We do this for 2 reasons:

    a)   it avoids using the Product Rule when differentiating (1) and

    b)   we wish to find $\frac{dh}{dt}$ only.

Note that it <u>is</u> valid to substitute for r <u>before</u> differentiating here because r is <u>always</u> equal to $\frac{h}{2}$. $\frac{h}{2}$ is not a special instant value for r.

$$V = \frac{1}{3}\pi r^2 h = \frac{1}{3}\pi \left(\frac{h}{2}\right)^2 h = \frac{\pi h^3}{12} \qquad (1)$$

Note that at the instant when $V = 486\pi$

$$\frac{\pi h^3}{12} = 486\pi$$

i.e. $\qquad h^3 = 5832$

i.e. $\qquad \underline{h = 18}$

Differentiate (1) with respect to time:

$$\frac{dV}{dt} = \frac{\pi h^2}{4} \frac{dh}{dt}$$

i.e. $\quad 40 = \frac{\pi (18)^2}{4} \frac{dh}{dt}$

$$\frac{40}{81\pi} = \frac{dh}{dt}$$

i.e. $\quad \frac{dh}{dt} = 0.1572 \ \text{cm/min}$

This means that the level of water is rising at a rate of 0.1572 cm per

minute.

Example

Question: A northbound ship leaves harbour at 12 noon with a speed of

7.5 knots and a westbound ship leaves the same harbour at 2 p.m. with

a speed of 8 knots. How fast are the ships separating at 4 p.m.?

Answer:

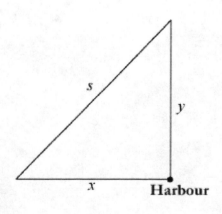

Let x nautical miles and y nautical

miles be the distances from the

harbour of the westbound and

northbound ships respectively. Let

s nautical miles be the distance

separating the two ships at an

arbitrary time t hours after noon.

We wish to find $\dfrac{ds}{dt}$ at 4 p.m. i.e. when $t=4$.

$$s^2 = x^2 + y^2 \qquad (1)$$

Differentiate with respect to time and cancel by 2 then:

$$s\frac{ds}{dt} = x\frac{dx}{dt} + y\frac{dy}{dt} \qquad (1)'$$

At 4 p.m. $x=16$ and $y=30$ and $s=34$ (Pythagoras).

$$\frac{dx}{dt}=8 \text{ and } \frac{dy}{dt}=7.5.$$

Substituting in (1)' we have

$$34\frac{ds}{dt} = 16(8)+30(7.5)$$

i.e. $\quad \dfrac{ds}{dt} = \dfrac{353}{34} = 10.38$ nautical miles per hour.

i.e. At 4 p.m. the distance separating the ships is increasing at approximately 10.38 nautical miles per hour.

Worksheet 1

1. If $A=\pi r^2$, find $\dfrac{dA}{dt}$ when $r=2$ and $\dfrac{dr}{dt}=3$.

2. If $A=2\pi rh$, find $\dfrac{dr}{dt}$ when $r=2$, $h=4$, $\dfrac{dA}{dt}=16\pi$ and $\dfrac{dh}{dt}=2$.

3. If $\dfrac{r}{3}=\dfrac{h-4}{h}$, find $\dfrac{dh}{dt}$ when $r=2$, $h=12$, and $\dfrac{dr}{dt}=\dfrac{1}{2}$.

4. If $A=\pi r\sqrt{h^2+r^2}$, find $\dfrac{dA}{dt}$ when $r=3$, $h=4$, $\dfrac{dr}{dt}=\dfrac{1}{5}$, and $\dfrac{dh}{dt}=\dfrac{1}{2}$.

5. A particle moves along the curve $y = \sqrt{x^2 + 1}$ in such a way that $\frac{dx}{dt} = 4$.

   Find $\frac{dy}{dt}$ when $x = 3$.

6. Two automobiles start from a point A at the same time. One travels west at 80 miles per hour; the other travels north at 45 miles per hour. How fast is the distance between them increasing 3 hours later?

7. A spherical balloon is being inflated at the rate of 12 in^3/sec. How fast is the radius r changing at the moment when $r = 2$ inches?

8. A particle is moving along the curve $y = x^2$ in such a way that when $x = \frac{1}{2}$,

   $\frac{dx}{dt} = 2$ ft/sec. Determine $\frac{dy}{dt}$ at that moment.

9. A ladder 15 feet tall leans against a vertical wall of a house. If the bottom of the ladder is pulled away horizontally from the house at constant speed of 4 ft/sec, how fast is the top of the ladder sliding down the wall when the bottom of the ladder is 9 feet from the wall?

10. A cone (point down) with a height of 10 inches and a radius of 2 inches is being filled with water at a constant rate of 2 in³/sec. Determine how fast the water surface is rising when the water depth is 6 inches.

11. A particle is moving along the graph of $y = \sqrt{x}$. At what point on the curve are the x-coordinate and the y-coordinate of the particle changing at the same rate?

12. A winch at the end of a dock is 10 feet above the level of the deck of a boat. A rope attached to the deck is being hauled in by the winch at a rate of 5 ft/sec. How fast is the boat being pulled toward the dock when 26 feet of rope are out?

13. The cross-section of a trough 6 feet long is an equilateral triangle with one vertex pointing down.

    a) Define a function which relates the volume of water in the trough to the depth of the water.

    b) If water is flowing into the trough at a rate of 10 ft³/sec, find the rate at which the depth of the water is increasing when the depth is 2.5 feet.

14. A streetlight is 15 feet above the sidewalk. A man 6 feet tall walks away from the light at a rate of 5 ft/sec.

    a) Determine a function relating the length of the man's shadow to his distance from the base of the streetlight.

    b) Determine the rate at which the man's shadow is lengthening at the moment that he is 20 feet from the base of the light.

Answers to Worksheet 1

1. 12π 2. 1   3. 6   4. $\dfrac{64}{25}\pi$ 5. $\dfrac{6\sqrt{10}}{5}$ 6. 91.79   7. $\dfrac{3}{4\pi}$ in/sec

8. 2 ft/sec   9. 3 ft/sec   10. $\dfrac{25}{18\pi}$ in/sec   11. $\left(\dfrac{1}{4},\dfrac{1}{2}\right)$ 12. $\dfrac{65}{12}$ ft/sec

13. a) $v(h)=2\sqrt{3}h^2$ b) $\dfrac{1}{\sqrt{3}}$ ft/sec   14. a) $s=\dfrac{2}{3}d$ b) $\dfrac{10}{3}$ ft/sec

Worksheet 2

RELATED RATES

1. A ladder 10 metres long is leaning against a wall when it begins to slide down the wall at a constant rate of 2 metres per second. Find the rate which the base of the ladder moves away from the wall when the top of the ladder is 6 metres above the ground.

2. At the instant when the radius of a cone is 3 inches the volume of the cone is changing at the rate of 9π cubic inches per minute. If the height is always 3 times the radius find the rate of change of the radius at that instant.

3. The volume of a cube is changing at the rate of 300 cubic inches per minute. Find the rate of change of the edge of the cube when the length of the edge is 10 inches. Find also the rate of change of the surface area of the cube when the surface area is a 150 square inches.

4. For each of the functions below find the rate at which y is increasing when $x = 4$ if x is increasing at a constant rate of 3 units per second.

   a) $y = \dfrac{x^2 - x - 2}{x - 3}$ 　　 b) $y = 4x^2 + \dfrac{1}{2x}$ 　　 c) $y = x^3 - 3x^2$ 　　 d) $y = 2x + \dfrac{1}{x^2}$

5. Sand is being poured onto a conical pile whose diameter is always twice its height. When the height is 10 metres its height is observed to be increasing at a rate of $\dfrac{1}{4}$ metre per minute. What is the rate of the flow of the sand?

6. The volume of a cylinder is increasing at the rate of 4π cubic cm per second. The radius of the base is increasing at the rate of 2 cm/sec. How fast is the height of the cylinder changing when the volume is 36π cubic cm and the radius of the base is 3 cm? Is the height increasing or decreasing?

7. Two parallel sides of a rectangle are being increased at a rate of 2 cm per second while the other 2 sides are being shortened so that the area remains constant at 50 square cm.

   a) What is the rate of change of the perimeter when the length of an increasing side is 5 cm?

   b) What are the dimensions at the instant when the perimeter stops decreasing?

Answers to Worksheet 2

1. $1\dfrac{1}{2}$

2. $\dfrac{1}{3}$

3. 1 inch/min,   240 square inches/min

4. a) -9   b) 95.9   c) 72   d) 5.9

5. 25π cubic metres per minute

6. $-\dfrac{44}{9}$, decreasing

7. a) -4   b) $5\sqrt{2}$ by $5\sqrt{2}$

Worksheet 3

1. If a, b and c are sides of a right triangle, where c is the hypotenuse and c is constant at 13 cm, given that a is increasing at a rate of 3 cm/s, find the rate of change of b in cm/s when $b = 5$ cm.

   (A) -20 cm/s   (B) 20 cm/s   (C) -1.25 cm/s

   (D) -7.2 cm/s   (E) 7.2 cm/s

2. A child is rolling some clay in the form of a cylinder. The radius of the cylinder decreases at a constant rate of 4 cm/min. At a certain instant when the radius is 8 cm, the height is 10 cm. The rate at which the height is increasing, in cm/min is: (A) -10   (B) 10   (C) 0   (D) 10π (E) 5

3. The sides of this rectangle increase in such a way that $\dfrac{dz}{dt} = 1.9$ and $\dfrac{dx}{dt} = 4\dfrac{dy}{dt}$. At the instant when $x = 8$ and $y = 6$, what is the value of $\dfrac{dx}{dt}$?

   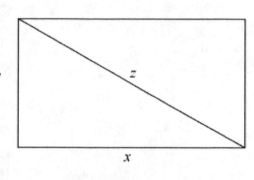

   (A) $\dfrac{1}{3}$ (B) 1   (C) 2   (D) $\sqrt{5}$ (E) 5

4. A spherical balloon with radius r inches is being inflated. The function V whose graph is sketched in the figure gives the volume of the balloon, $V(t)$, measured in cubic inches after t seconds. At what approximate rate is the radius of the balloon changing after 4 seconds?

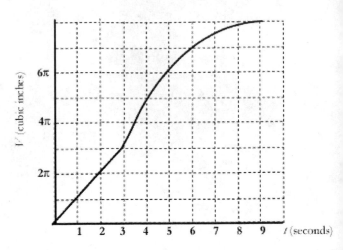

(A) 5π (B) .15536   (C) .0495   (D) 4.817   (E) .488

5. Children are rolling a ball of snow to form a snowman. The radius of the ball is changing according to the law $r = \dfrac{2t^2}{t^2+1}$, where t is in minutes and r is in metres.

a) $\dfrac{dV}{dt}$ after 1 minute is:

    (A) $\dfrac{4\pi}{3}$ (B) 8π (C) 4π (D) $\dfrac{8}{25}$ (E) 1

b) What is the limit to the volume of snow as time goes on indefinitely?

    (A) 16π (B) $\dfrac{8\pi}{3}$ (C) $\dfrac{4\pi}{3}$ (D) $\dfrac{32\pi}{3}$ (E) 32π

6. Sand is dumped on a pile in such a way that it forms a cone whose base radius is always 3 times its height. The function V whose graph is sketched in the figure gives the volume of the conical sand pile, $V(t)$.

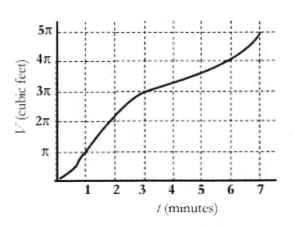

At what approximate rate is the radius of the base changing after 6 minutes?

(A) .0438   (B) .138   (C) .0146   (D) .046   (E) $\dfrac{1}{87\pi}$

7. The function V whose graph is sketched below gives the volume of air $V(t)$ that a man has blown into a sperical balloon after t seconds. Approximately how rapidly is the radius changing after 6 seconds?

(A) $\dfrac{1}{5}$ (B) $\dfrac{\pi}{5}$ (C) 3   (D) $\dfrac{\pi}{3}$ (E) 2.813

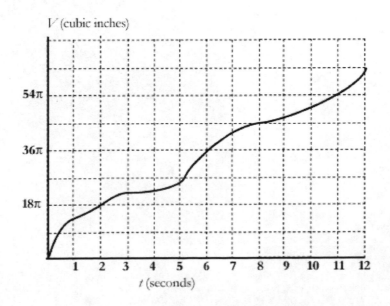

<u>Answers to Worksheet 3</u>

1. D

2. B

3. C

4. B

5. a) C

   b) D

6. B

7. A

Example (in 3-dimensional space)

Question: A man is running over a bridge at a rate of 5 metres per second while a

boat passes under the bridge and immediately below him at a rate of

1 metre per second. The boat's course is at right angles to the man's

and 6 metres below it. How fast is the distance the between the man

and the boat separating 2 seconds later?

Answer:

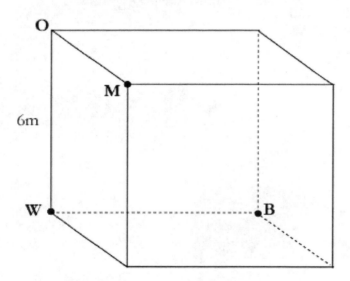

In the diagram let M be the position of the man, let B be the position

of the boat. *OM* represents the distance the man runs. *WB*

represents the distance the boat goes along the river.

OW is 6 m i.e. the height of the bridge above the water.

Let *OM* be x, let *WB* be y and let *MB* be s, the distance separating

the man and the boat.

$$s^2 = x^2 + 6^2 + y^2 \qquad (1)$$

After two seconds, $x = 10$ m, $y = 2$ m, and hence $s = \sqrt{140}$.

Differentiate (1) with respect to time and cancel by 2.

Then $\quad s\dfrac{ds}{dt} = x\dfrac{dx}{dt} + y\dfrac{dy}{dt}$

After 2 seconds we have

$$\sqrt{140}\,\frac{ds}{dt} = 10(5) + 2(1)$$

$\therefore \qquad \dfrac{ds}{dt} = \dfrac{52}{\sqrt{140}} = 4.395 \ \text{(approx.)}$

$\therefore \qquad$ The distance between the man and the boat is increasing by

4.395 m/sec.

Example

Question: In the diagram shown the corridors are 4 m and 5 m wide. A cat C is moving to the right at 10 m/sec. At the instant when a dog D is 12 metres from A, how fast must the dog run in order to keep the cat in sight?

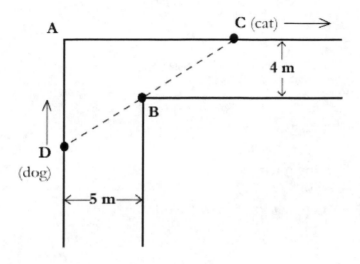

Answer: Let $AD = y$ and $AC = x$. We wish to find $\dfrac{dy}{dt}$ when $y = 12$ and $\dfrac{dx}{dt} = 10$.

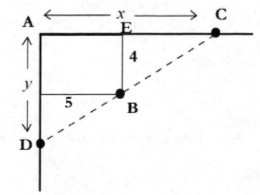

In the diagram,

$$\Delta DAC \parallel \Delta BEC$$

$$\therefore \quad \frac{DA}{BE} = \frac{AC}{EC}$$

$$\therefore \quad \frac{y}{4} = \frac{x}{x-5} \, .$$

$$\therefore \quad y = \frac{4x}{x-5} \qquad (1)$$

Differentiate (1) with respect to time:

$$\frac{dy}{dt} = \frac{(x-5)4\frac{dx}{dt} - 4x\frac{dx}{dt}}{(x-5)^2}$$

At the special instant when $y = 12$, $x = 7.5$ from (1)

and $\quad \frac{dx}{dt} = 10$. We have:

$$\frac{dy}{dt} = \frac{(7.5-5)4(10) - 4(7.5)(10)}{((7.5)-5)^2}$$

$$= -32 \text{ m/sec}$$

$\therefore \quad$ The dog must move at 32 m/sec to keep the cat in sight. (The

dog is going to be disappointed.)

Worksheet 4

<u>RELATED RATES</u>

1. Sand is being dropped on the ground at a steady rate of 9π cubic cm per

   second and forms a conical pile whose height remains equal to the radius of

   the base. At what rate is the height increasing after 8 seconds?

2. The area of a circular ink blot starts from zero and grows at a constant rate of

   4π square cm per minute. Find the rate at which the radius is increasing:

   a) after 3 minutes.   b) when the radius is 1 cm.

3. A basin has the shape of an inverted cone with altitude 6 cm and radius at the top of 9 cm. Water runs in at the constant rate of 3 cubic cm per second. At the instant when the water is 4 cm deep find:

a) the rate at which the level of water is rising.

b) the rate of increase of the exposed surface area of the water.

4. A barge whose deck is 5 metres below the level of a dock is being drawn in by means of a cable attached to the dock and passing through a ring on the dock. When the barge is 12 metres away and approaching the dock at $\frac{3}{4}$ metre per minute, how fast is the cable being pulled in?

5. Two aeroplanes fly eastwards on parallel courses 12 miles apart. One flies at 240 m.p.h. and the other at 300 m.p.h. How fast is the distance between them changing when the slower plane is 5 miles farther east than the faster plane?

6. A ladder rests against a vertical pole. The foot of the ladder is sliding away from the pole along horizontal ground. Find the inclination of the ladder to the horizontal at the instant when the top of the ladder is moving 3 times as fast as the foot of the ladder.

7. A cone of fixed height 12 inches is changing in shape through the change in the radius of the base. What rate of increase of the radius will make the lateral surface area of the cone increase at the rate of 10π square inches per minute when the radius of the base is 5 inches?

8. A northbound ship leaves harbour at 10 a.m. with a speed of 12 knots. A westbound ship leaves harbour at 10:30 a.m. with a speed of 16 knots. How fast are the ships separating at 11:30 a.m.?

9. At a certain instant, Submarine A is on the surface of the ocean and submarine B is 700 m due north of A and 100 m below the surface of the water. B is descending vertically at a constant rate of 2 m/s and A is moving north at a constant speed of 6 m/s.

   a) At what rate is the distance between A and B changing at this instant?

   b) At what rate will the distance between them change when A is directly above B?

   c) Where will A be when they are closest together?

Answers to Worksheet 4

1. $\frac{1}{4}$ cm per second

2. a) $\frac{1}{\sqrt{3}}$ b) 2

3. a) $\frac{1}{12\pi}$ b) $\frac{3}{2}$

4. $\frac{9}{13}$

5. -23.08

6. 18.4°

7. 0.67

8. 19.6 knots

9. a) $-4\sqrt{2}$ m/sec   b) 2 m/sec

   c) 100 m from being directly above B

Worksheet 5

REVIEW OF RELATED RATES

1. A balloon is being filled with air at a rate of 3 cubic metres per minute. Find the rate at which the radius is increasing when the volume is 36π cubic metres.

2. The slant height of a cone remains constant at 6 metres. If the radius is increasing at 5 metres per second how is the volume changing when the radius is 3 metres? What is the maximum volume?

3. At 12 noon ship A is 60 miles west of point P steaming east at 15 knots and ship B is 36 miles south of P steaming north at 10 knots. If the ships continue their courses and speeds,

   a) how is the distance between them changing at 2 p.m.?

   b) at what time are the ships closest?

4. Plaster is being applied uniformly to a cannon-ball at the rate of 24π cubic inches per minute. When the radius of the ball and plaster is 6 inches at what rate is the surface area increasing?

5. Two points, A and B, are 275 ft apart. At a given instant, a balloon is released at B and rises vertically at a constant rate of 2.5 ft/sec, and, at the same instant, a cat starts running from A to B at a constant rate of 5 ft/sec.

   a) After 40 seconds, is the distance between the cat and the balloon decreasing or increasing? At what rate?

b) Describe what is happening to the distance and the balloon at $t = 50$ seconds.

6. The cross section of a trough is a trapezoid with the lower base 1 metre, the upper base 2 metres and the depth 1 metre. The length of the trough is 6 metres. If water is poured in at the rate of 12 cubic metres per minute, at what rate is the water rising when the depth of water is $\dfrac{1}{3}$ metre?

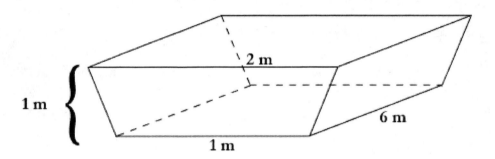

Answers to Worksheet 5

1. $\dfrac{1}{12\pi}$

2. a) $25\sqrt{3}\,\pi$ b) $16\sqrt{3}\,\pi$

3. a) 17.94 m.p.h.   b) 3:53 p.m.

4. 8π square inches per minute

5. a) decreasing at 1 ft/sec   b) increasing at 1.471 ft/sec

6. 1.5 metres per minute

Worksheet 6

Related Rates

1. A light is at the top of a pole 80 feet high. A ball is dropped at the same height from a point 20 feet away from the light. A wall 80 feet high, 60 feet away from the light is built. Assuming the ball falls according to the Newtonian Law $s = 16t^2$ where s is the distance in feet and t is the time in seconds, find:

   a) how fast the shadow of the ball is moving on the wall after 1 second.

   b) how fast the shadow is moving along the ground after 2 seconds.

2. Two poles are 24 metres and 30 metres high and 20 metres apart. A slack wire joins the tops of the poles and is 32 metres long. A cable car is moving along the wire at 5 metres per second away from the shorter pole. When the car is 12 metres horizontally from the shorter pole, it is 15 metres high and the length of the wire to the shorter pole is 15 metres. At this instant, find:

   a) how fast the car is moving horizontally.

   b) how fast the car is moving vertically.

3. An isosceles triangle is inscribed inside a circle of radius 12 metres. Its height is increasing at a rate of 2 m/sec. At the instant when the height of the triangle is 18 m, at what rate is:

   a) its area changing?

   b) its perimeter changing?

4. Water is being withdrawn from a conical reservoir, 3 metres radius and 10 metres deep at 4 cubic metres per minute. a) How fast is the surface of the water falling when the depth of water is 6 metres?

b) How fast is the radius of this surface diminishing at this instant?

5. The upper chamber of an hour-glass is a cone of radius 3 inches and height 10 inches and, if full, it requires exactly one hour to empty. Assuming that the sand falls through the aperture at a constant rate, how fast is the level falling when:

a) the depth of the sand is 6 inches?

b) $52\frac{1}{2}$ minutes have elapsed from the time when the hour-glass was full in

the upper chamber?

6. A man 6 feet tall walks along a walkway which is 30 feet from a the base of a lamp which is 126 feet tall. The man walks at a constant rate of 3 feet per second. How fast is the length of his shadow changing when he is 40 feet along the walkway past the closest point to the lamp?

126 feet

Answers to Worksheet 6

1. a) 96   b) 25   2. a) 7.5   b) $\frac{5}{3}$ 3. a) 0   b) 0

4. a) 0.393   b) 0.118   5. a) 9.26 inches/hour   b) 13.33 inches/hour

6. $\frac{3}{25}$ feet/second

Worksheet 7

1. A tank is leaking water. $V(t)$ represents the number of gallons of water in the tank after t minutes and $V(t) = 20(30-t)^2$.

   a) Find how fast the water is leaking from the tank after 10 minutes?

   b) Find the specific time $t = k$ if the average rate at which the water leaks from the tank between $t = k$ and $t = 2k$ is 280 gallons per minute.

2. A conical shape funnel 8 cm across at the top and 12 cm deep is leaking water at the rate of 2 cubic cm per minute.

   a) At what rate is the water level dropping at the instant when the water is 6 cm deep?

   b) At what rate is the water level dropping at the instant when the funnel is half full (by volume)?

3. A southbound ship leaves harbour at 12 noon with a speed of 10 knots. A westbound ship leaves harbour at 1 p.m. with a speed of 20 knots. How fast are the ships separating at 2 p.m.?

4. The slant height of a cone remains fixed at 10 metres. If the height of the cone is changing at a rate of 2 metres per minute, how fast is the volume changing at the instant when the radius of the base of the cone is 8 metres?

5. A cylinder of circular cross-section, initially has a base radius of 6 cm and a height of 10 cm. The radius is increasing at a constant rate of 1 cm/sec and the height is decreasing at a constant rate of 2 cm/sec.

a) Find the initial rate of change of the volume of the cylinder.

b) Find the rate of change of the volume of the cylinder after 1 second.

c) After how many seconds does the volume of the cylinder achieve its maximum value. What is that maximum volume?

Answers to Worksheet 7

1. a) 800 gallons/minute

b) $24\dfrac{2}{3}$ seconds

2. a) $\dfrac{1}{2\pi}$ b) 0.063

3. $15\sqrt{2}$ (21.21)

4. Volume is decreasing at $\dfrac{16\pi}{3}$ metres/min

5. a) Volume is increasing at 48π cm/sec

b) Volume is increasing at 14π cm/sec

c) After $\dfrac{4}{3}$ seconds. Max volume = 1238.95

Worksheet 8

1. The length L of a rectangle is decreasing at the rate of 2 cm/sec while the width W is increasing at the rate of 2 cm/sec. When $L = 12$ and $W = 2$, find the rate of change of :

   a) the area

   b) the perimeter

   c) the lengths of the diagonals

2. An aircraft is 100 km east of a radar beacon and is travelling west at 600 km/h. At the same instant, a helicopter flying at the same altitude is 48 km south of the radar beacon and is travelling south at 300 km/h. How fast is the distance between the aircraft and the helicopter changing after one minute.

3. The cross section of a water trough is an equilateral triangle. The trough is 5 metres long and 25 cm deep. The water is flowing in at a rate of 0.25 cubic metres per minute. How fast is the water level rising when the water level is 10 cm deep?

4. A light is at the top of a 30 m pole. A ball is dropped from the same height from a point 10 m from the light. The height of the ball (in metres) t seconds after it has been dropped is given by the formula $h = 30 - 5t^2$. How fast is the shadow of the ball moving along the ground after 1 second?

Answers to Worksheet 8

1. a) increasing at 20 cm/sec

   b) 0

   c) decreasing at 1.644 cm/sec

2. The distance is decreasing at 364.94 km/h

3. The water level is rising at 0.433 m/min

4. 120 m/sec

CHAPTER 9

Trigonometry

The concept upon which differentiation of trigonometric functions depends is based on the fact that

$$\lim_{\theta \to 0} \frac{\sin \theta}{\theta} = 1 \ .$$

It should be remembered that, at the Calculus level, when we talk of trigonometric functions, θ is measured in radians.

<u>Theorem</u> $\lim_{\theta \to 0} \dfrac{\sin \theta}{\theta} = 1$

<u>Proof</u>

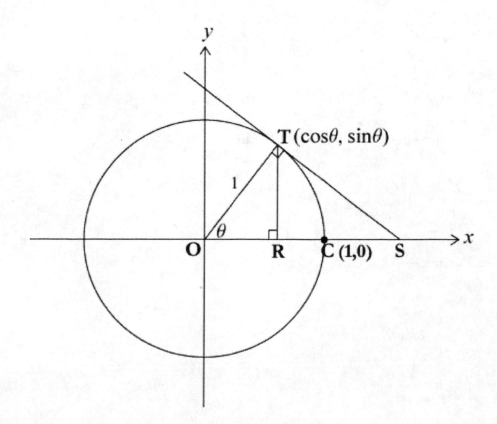

Consider the unit circle shown. OT is the radius, TS is a tangent and TR is perpendicular to OS. $\overset{\frown}{TC}$ is an arc of the unit circle as shown. Let $\angle TOC = \theta$.

Note that $OT = OC = 1$, $OR = \cos\theta$, $TR = \sin\theta$ and $TS = \tan\theta$.

Now   Area $\triangle OTR$ < Area sector $\overset{\frown}{OTC}$ < Area $\triangle OTS$

$\therefore \qquad \dfrac{1}{2}\cos\theta\sin\theta < \dfrac{1}{2}\theta\cdot 1^2 < \dfrac{1}{2}\cdot 1\cdot\tan\theta$

$\therefore \qquad \cos\theta\sin\theta < \theta < \tan\theta$

$\therefore \qquad \cos\theta\sin\theta < \theta < \dfrac{\sin\theta}{\cos\theta}$

We will be examining what happens as $\theta \to 0$ so here it is appropriate to consider θ as acute and hence $\sin\theta$ is positive.

$\therefore \qquad \cos\theta < \dfrac{\theta}{\sin\theta} < \dfrac{1}{\cos\theta}$

In the limiting case as $\theta \to 0$ we have

$$\lim_{\theta\to 0}\cos\theta \ \le\ \lim_{\theta\to 0}\dfrac{\theta}{\sin\theta} \ \le\ \lim_{\theta\to 0}\dfrac{1}{\cos\theta} \quad \textbf{✱}$$

$\therefore \qquad 1 \le \lim_{\theta\to 0}\dfrac{\theta}{\sin\theta} \le 1$

i.e. $\qquad \lim_{\theta\to 0}\dfrac{\theta}{\sin\theta}$ is "sandwiched" between 1 and 1

$\therefore \qquad \lim_{\theta\to 0}\dfrac{\theta}{\sin\theta} = 1$

$\therefore \qquad \lim_{\theta\to 0}\dfrac{\sin\theta}{\theta} = 1 \quad$ (since $\lim\dfrac{f(x)}{g(x)} = \dfrac{1}{\lim\dfrac{g(x)}{f(x)}}$).

Note that in the inequality at ✱ the equality symbols are added since for example

when $0 < x < 1$, $x^3 < x^2$

but $\lim\limits_{x \to 0} x^3 = \lim\limits_{x \to 0} x^2 \quad (= 0)$

From a graphical point of view also note that both $y = \dfrac{x+2}{x+1}$ and $y = \dfrac{x+3}{x+1}$

have $y = 1$ as an asymptote when x gets large.

For all positive values $\dfrac{x+2}{x+1} < \dfrac{x+3}{x+1}$

But $\lim\limits_{x \to \infty} \dfrac{x+2}{x+1} = \lim\limits_{x \to \infty} \dfrac{x+3}{x+1} \quad (= 1)$

Example

a) *Find:* $\lim\limits_{x \to 0} \dfrac{\sin x}{3x}$.

  Answer: $\lim\limits_{x \to 0} \dfrac{\sin x}{3x} = \lim\limits_{x \to 0} \dfrac{1}{3} \dfrac{\sin x}{x}$

          $= \dfrac{1}{3} \cdot 1 = \dfrac{1}{3}$

b) *Find:* $\lim\limits_{x \to 0} \dfrac{\sin 3x}{x}$.

  Answer: $\lim\limits_{x \to 0} \dfrac{\sin 3x}{x} = \lim\limits_{x \to 0} \dfrac{3 \sin 3x}{3x}$

          $= 3 \lim\limits_{x \to 0} \dfrac{\sin 3x}{3x}$

$$= 3 \lim_{u \to 0} \frac{\sin u}{u} \qquad (\text{where } u = 3x)$$

$$= 3 \cdot 1 = \underline{\underline{3}}.$$

c) *Find*: $\lim_{x \to 0} \dfrac{1 - \cos x}{x^2}$

Answer: $\lim_{x \to 0} \dfrac{1 - \cos x}{x^2} = \lim_{x \to 0} \dfrac{(1 - \cos x)(1 + \cos x)}{x^2 (1 + \cos x)}$

$$= \lim_{x \to 0} \frac{\left(1 - \cos^2 x\right)}{x^2 (1 + \cos x)} = \lim_{x \to 0} \frac{\sin^2 x}{x^2 (1 + \cos x)}$$

$$= \lim_{x \to 0} \frac{\sin x}{x} \cdot \lim_{x \to 0} \frac{\sin x}{x} \cdot \lim_{x \to 0} \frac{1}{1 + \cos x}$$

$$= 1 \cdot 1 \cdot \frac{1}{2} = \frac{1}{2}$$

Worksheet 1

Find the following limits (if they exist):

1. $\lim_{x \to -1} \dfrac{\sin(x+1)}{(x+1)}$

2. $\lim_{x \to 0} \dfrac{\sin(x+1)}{x+1}$

3. $\lim_{x \to 0} \dfrac{\sin^2 x}{x}$

4. $\lim_{x \to 0} \dfrac{\sin^2 x}{x^3}$

5. $\lim_{x \to 0} \dfrac{\sin 2x}{3x}$

6. $\lim_{x \to 0} \dfrac{\tan x}{x}$

7. $\lim_{x \to 0} \dfrac{1 - \cos x}{\sin^2 x}$

8. $\lim_{x \to 0} x \sin \dfrac{1}{x}$

9. $\lim_{\theta \to 0} \dfrac{\sin 5\theta + \sin 7\theta}{6\theta}$

10. $\lim_{x \to 0} \dfrac{x^2}{1 - \cos x}$

11. $\lim_{x \to \infty} x \sin \dfrac{1}{x}$

12. $\lim_{x \to \frac{\pi}{2}} \dfrac{\cos 3x + \cos 5x}{4x - 2\pi}$

Answers to Worksheet 1

1. 1   2. $\sin 1$ 3. 0   4. no limit   5. $\dfrac{2}{3}$ 6. 1   7. $\dfrac{1}{2}$ 8. 0   9. 2   10. 2   11. 1   12. $-\dfrac{1}{2}$

To Investigate the Derivative of *sin x*

Let $f(x) = \sin x$.

By definition, $f'(x) = \lim\limits_{h \to 0} \dfrac{f(x+h) - f(x)}{h}$

$$= \lim\limits_{h \to 0} \frac{\sin(x+h) - \sin x}{h}$$

$$= \lim\limits_{h \to 0} \frac{2\cos\left(\dfrac{2x+h}{2}\right)\sin\left(\dfrac{h}{2}\right)}{h} \quad \text{(Trig identity)}$$

$$= \lim\limits_{h \to 0} \cos\left(\frac{2x+h}{2}\right)\lim\limits_{h \to 0}\frac{\sin\dfrac{h}{2}}{\left(\dfrac{h}{2}\right)}$$

$$= \cos x \cdot 1 = \cos x$$

$$\therefore \quad \boxed{D(\sin x) = \cos x}$$

To Investigate D(*cos x*)

$$D(\cos x) = D\left(\sin\left(\frac{\pi}{2} - x\right)\right)$$

$$= \cos\left(\frac{\pi}{2} - x\right) \cdot (-1)$$

$$= -\sin x$$

$$\therefore \quad \boxed{D(\cos x) = -\sin x}$$

Similarly, $D(\tan x) = D\left(\dfrac{\sin x}{\cos x}\right)$

$$= \frac{\cos x \cdot \cos x - \sin x(-\sin x)}{(\cos x)^2} \qquad \text{(Quotient Rule)}$$

$$= \frac{\cos^2 x + \sin^2 x}{\cos^2 x}$$

$$= \frac{1}{\cos^2 x} = \sec^2 x$$

$\therefore \qquad \boxed{D(\tan x) = \sec^2 x}$

Example

Find $\dfrac{dy}{dx}$ for $y = \sin^2 3x$. Remember that $\sin^2 3x$ means $(\sin 3x)^2$

$\therefore \qquad \dfrac{dy}{dx} = 2(\sin 3x) \cdot \cos 3x \cdot 3 \qquad$ Chain Rule

$$= 6 \sin 3x \cos 3x$$

Example

Find an equation of the tangent to $\cos x \sin y + \cos y = x + 1$ at the

point $\left(0, \dfrac{\pi}{2}\right)$.

Differentiate with respect to x.

$$\sin y(-\sin x) + \cos x \left(\cos y \frac{dy}{dx}\right) - \sin y \frac{dy}{dx} = 1$$

At $\left(0, \dfrac{\pi}{2}\right)$

$$0 + 0 - \frac{dy}{dx} = 1$$

$\therefore$ slope of the tangent at $\left(0, \dfrac{\pi}{2}\right)$ is -1.

$\therefore$ an equation of the tangent at $\left(0, \dfrac{\pi}{2}\right)$ is

$$y - \frac{\pi}{2} = -1(x - 0)$$

i.e. $x + y = \dfrac{\pi}{2}.$

When differentiating other trig functions the following results are obtained:

$$\left.\begin{array}{l} D(\sin x) = \cos x \\ D(\cos x) = -\sin x \end{array}\right\}$$

$$\left.\begin{array}{l} D(\tan x) = \sec^2 x \\ D(\sec x) = \sec x \tan x \end{array}\right\}$$

$$\left.\begin{array}{l} D(\cot x) = -\operatorname{cosec}^2 x \\ D(\operatorname{cosec} x) = -\operatorname{cosec} x \cot x \end{array}\right\}$$

Note that, for memory purposes, in trig formulae and differentiation

$\sin x$ is associated with $\cos x$

$\tan x$ is associated with $\sec x$

$\cot x$ is associated with $\operatorname{cosec} x$

Unlike most textbooks, note that $\csc x$ is written $\operatorname{cosec} x$ for the following reason.

It is important to understand that the prefix co... is an abbreviation for complement.

e.g. $\operatorname{cosec} x \equiv \sec\left(\dfrac{\pi}{2} - x\right)$

$\cot x \equiv \tan\left(\dfrac{\pi}{2} - x\right)$

$\cos x \equiv \sin\left(\dfrac{\pi}{2} - x\right)$

Hence, when differentiating a co... function a negative coefficient will arise from the Chain Rule.

i.e. $D(\text{co.....}) = -.....$

The authors lament the decline in usage of $\operatorname{cosec} x$ and almost universal usage of $\csc x$.

Example

Show that $y = \tan x$ and $y = \sin x$ are tangent to one another.

It suffices to show that, at a point of intersection, their slopes are equal.

At a point of intersection $\tan x = \sin x$

$$\text{and} \quad \frac{\sin x}{\cos x} = \sin x.$$

Hence $x = 0$ yields one point of intersection, i.e. the graphs intersect at $(0,0)$.

$D(\tan x) = \sec x$. At $(0,0)$ slope of $y = \tan x$ is $\sec^2 0$ which is 1.

$D(\sin x) = \cos x$. At $(0,0)$ slope of $y = \sin x$ is $\cos 0$ which is also 1.

$\therefore$ The two graphs are tangent to one another at the origin.

Incidentally, this can be deduced by non-calculus methods by noting that the

solution(s) to $\tan x = \sin x$ lead to $\dfrac{\sin x}{\cos x} = \sin x$

i.e. $\sin x = \sin x \cos x$ (assuming $\cos x \neq 0$)

i.e. $0 = \sin x \cos x - \sin x$

$0 = \sin x (\cos x - 1)$

which has a double root when $x = 0$ and hence the graphs are tangent.

———————————

Example

Find the derivative of Arctan$x$.

Let $y = $ Arctan$x$ where domain of this relation is $\{x : -\infty < x < +\infty\}$ and the range is $\left\{ y : -\dfrac{\pi}{2} < y < \dfrac{\pi}{2} \right\}$.

Then $\tan y = x$

Differentiate with respect to x.

$$\sec^2 y \frac{dy}{dx} = 1$$

$$\therefore \quad \frac{dy}{dx} = \frac{1}{\sec^2 y} = \frac{1}{1 + \tan^2 y} = \frac{1}{1 + x^2}$$

$$\therefore \quad D(\text{Arctan}x) = \frac{1}{1 + x^2}.$$

When differentiating Arcsin$x$ or Arccos$x$ difficulties arise and conventional wisdom is that domain of Arcsin$x$ and Arccos$x$ is $\{x : -1 \leq x \leq 1\}$ and the range of these relations is the set of values (radians) closest in value to zero choosing positive values when necessary. For example, range of the Arcsin relation is $\left\{ y : -\dfrac{\pi}{2} \leq y \leq \dfrac{\pi}{2} \right\}$ and the range of the Arccos relation is $\{y : 0 \leq y \leq \pi\}$.

Worksheet 2

1. Differentiate with respect to x:

   a) $\sin x \cos x$ b) $\sec^2 x$ c) $\tan^2 x$ d) $\tan 2x$

   e) $\csc\left(x^2 - 1\right)$ f) $\dfrac{\sin x}{\cos 2x}$ g) $\cos\left(\sin x\right)$ h) $3\cot^2 5x$

   i) $\csc^2 3x$ j) $\sin\left(x^2\right) + \sin^2 x$ k) $\sec^2\left(3x^2\right)$ l) $\sec x \cot x$

   m) $\tan^3 x$

2. Find the equation of the tangent to $y = 2\sec x$ at $\left(\dfrac{\pi}{3}, 4\right)$.

3. At what angle does $y = \tan x$ pass through the origin?

4. Find $\dfrac{dy}{dx}$ if $y + \cos y = x^2 + \tan x$.

5. Find a relative minimum point on the graph of $y = x - \sin 2x$.

6. Explain why $D\left(\sec^2 x\right) = D\left(\tan^2 x\right)$ but $D\left(\tan x\right) \neq D\left(\sec x\right)$.

7. Find the slope of $2\sin\left(x + y\right) + \cos x = 1 + y$ at $\left(0, 0\right)$.

8. Find a relative maximum point on the graph of $y = x - 2\cos x$.

9. Find the equation of the tangent to $y = x - \cos x$ at the point where $x = 0$.

   Does the tangent intersect the curve again? If so, where?

10. Graph $y = x - 2\sin x$ showing a maximum point, a minimum point and an inflection point.

11. Show that $y = x + \sin x$ has no relative maximum or relative minimum.

*12. Investigate whether $y = \dfrac{x}{\sin x}$ has a relative maximum or relative minimum.

Answers to Worksheet 2

1. a) $\cos^2 x - \sin^2 x$ b) $2\sec^2 x \tan x$ c) $2\tan x \sec^2 x$ d) $2\sec^2 2x$

 e) $-2x\operatorname{cosec}(x^2 - 1)\cot(x^2 - 1)$ f) $\dfrac{\cos 2x \cos x + 2\sin x \sin 2x}{\cos^2 2x}$

 g) $-\sin(\sin x)$ times $\cos x$ h) $-30\cot 5x \operatorname{cosec}^2 5x$

 i) $-6\operatorname{cosec}^2 3x \cot 3x$ j) $2x\cos(x^2) + 2\sin x \cos x$

 k) $12x\sec^2(3x^2)\tan(3x^2)$ l) $-\operatorname{cosec} x \cot x$

 m) $3\tan^2 x \sec^2 x$

2. $y - 4 = 4\sqrt{3}\left(x - \dfrac{\pi}{3}\right)$ 9. $y = x - 1$

3. $45°$ Yes at $(2n\pi, 2n\pi - 1)$ for all integers n

4. $\dfrac{2x + \sec^2 x}{1 - \sin y}$ 10. $\left(\dfrac{\pi}{3}, \dfrac{\pi}{3} - \sqrt{3}\right)$ minimum

5. $\left(\dfrac{\pi}{6}, \dfrac{\pi}{6} - \dfrac{\sqrt{3}}{2}\right)$ $\left(\dfrac{5\pi}{3}, \dfrac{5\pi}{3} + \sqrt{3}\right)$ maximum

7. -2 $(0,0)$ and (π,π) inflection

8. $\left(\dfrac{7\pi}{6}, \dfrac{7\pi}{6} + \sqrt{3}\right)$ 12. graph has relative max/min when

 $\tan x = x$ e.g. a relative max at $(4.493, -4.603)$

Trigonometry: Max/Min and Related Rate Problems

Example

Question: If the angle of elevation of the sun is 60° and is increasing at a rate of 15° per hour, how fast is the shadow cast by a 30 metre pole decreasing (or increasing) in length?

Answer:

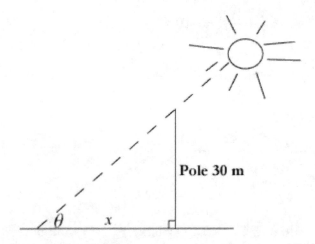

This is a related rate problem. We know $\dfrac{d\theta}{dt} = 15°$ per hour and we

wish to find $\dfrac{dx}{dt}$ when $\theta = 60°$. Firstly, it must be remembered that we must

use <u>radians</u>. i.e. $\dfrac{d\theta}{dt} = \dfrac{\pi}{12}$ radians/hour.

$\because \tan\theta = \dfrac{30}{x}$

$\therefore x = 30\cot\theta$

Differentiate with respect to time

$$\frac{dx}{dt} = -30\text{cosec}^2\theta\frac{d\theta}{dt}$$

At the special instant when $\theta = \dfrac{\pi}{3}$ and $\dfrac{d\theta}{dt} = \dfrac{\pi}{12}$

We have $\dfrac{dx}{dt} = -30\text{cosec}^2\left(\dfrac{\pi}{3}\right)\dfrac{\pi}{12}$

$$= -\frac{10\pi}{3} = -10.47 \text{ (approx.)}$$

∴   the shadow is decreasing in length at a rate of 10.47 metres per

hour.

_____

Example

Question: An illuminated advertising billboard 10 m tall stands on top of a cliff

12 m high. How far from the foot of the cliff should a man stand in

order for the sign to subtend the largest possible angle at his eyes

which are 2 metres above the ground? How large is the maximum

angle?

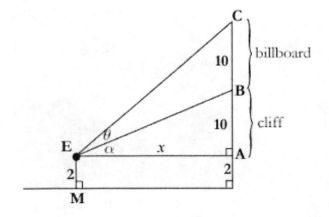

Answer: $BC = 10$ m is the billboard. $AB = 10$ m is the part of the cliff above the level of the man's eyes. Let x be the distance from the man's eyes to the cliff. Let θ be the angle subtended at the man's eyes by the billboard. We wish to find x to maximize θ.

Let $EA = x$ and let $\measuredangle BEA = \alpha$.

In $\triangle BEA$ $\tan \alpha = \dfrac{10}{x}$ (1)

In $\triangle CEA$ $\tan(\theta + \alpha) = \dfrac{20}{x}$ (2)

$\therefore$ $\dfrac{\tan \theta + \tan \alpha}{1 - \tan \theta \tan \alpha} = \dfrac{20}{x}$ (2) (trig identity)

Substituting for $\tan \alpha$ from (1) into (2) we get

$$\frac{\tan \theta + \dfrac{10}{x}}{1 - \tan \theta \left(\dfrac{10}{x} \right)} = \frac{20}{x}$$

i.e. $\tan \theta = \dfrac{10x}{x^2 + 200}$ (2)

Differentiate (2) with respect to x.

$$\sec^2 \frac{d\theta}{dx} = \frac{\left(x^2 + 200\right)10 - 10x(2x)}{\left(x^2 + 200\right)^2}$$

If is a maximum then $\dfrac{d\theta}{dx} = 0$

$\therefore \left(x^2 + 200\right)10 - 10x(2x) = 0,$ i.e. $x^2 = 200 \Rightarrow x \doteq 14.14$

From a graphical and analytical point of view it is not clear that this value of x yields a <u>maximum</u> value for θ. However common sense dictates that θ can get as small as one likes simply by standing farther and farther away from the angle so no minimum angle exists.

Reference to a graph of $\theta = \arctan \dfrac{10x}{x^2 + 200}$ justifies this, also. Further justification for the fact that $x = \sqrt{200}$ yields a <u>maximum</u> angle θ can be obtained by showing that $\dfrac{d^2\theta}{dx^2}$ is negative which is left as an exercise for the reader.

Example

Question: A lighthouse has a revolving light which rotates at a rate of 4 revolutions per minute. The lighthouse is situated on rocks 2 km out to sea from a straight coast line. Find how fast the spot of light from the lighthouse is moving along the coastline as it passes through the point on the coastline nearest to the lighthouse.

Answer:

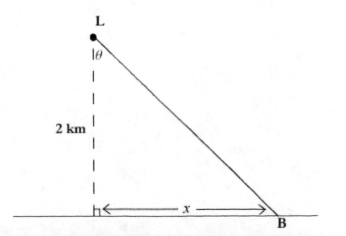

Let x and θ be as shown in the diagram.

We know $\dfrac{d\theta}{dt} = 4$ revolutions/minute $= 8\pi$ radians/minute.

We wish to find $\dfrac{dx}{dt}$ at the instant when $\theta = 0$.

$$\tan\theta = \frac{x}{2}$$

Differentiate with respect to time:

$$\sec^2\theta \frac{d\theta}{dt} = \frac{1}{2}\frac{dx}{dt}$$

When $\theta = 0$ we have: $1^2 \cdot 8\pi = \dfrac{1}{2}\dfrac{dx}{dt}$

$$\frac{dx}{dt} = 16\pi \text{ km/minute}$$

Worksheet 3

1. Find: $\displaystyle\lim_{x \to 0} \frac{\sin 4x}{x}$.

   (A) -4   (B) 0   (C) $\dfrac{1}{4}$ (D) 4   (E) ∞

2. If $y = \cos^2(2x)$, then $\dfrac{dy}{dx} =$

   (A) $2\cos 2x \sin 2x$ (B) $-4\sin 2x \cos 2x$ (C) $2\cos 2x$

   (D) $-2\cos 2x$ (E) $4\cos 2x$

3. $\displaystyle\lim_{h \to 0}\left(\frac{\cos(x+h) - \cos x}{h}\right) =$

   (A) $\sin x$ (B) $-\sin x$ (C) $\cos x$ (D) $-\cos x$ (E) does not exist

4. Evaluate: $\displaystyle\lim_{x \to 0} \frac{1 - \cos^2 x}{x \sin 2x}$

   (A) 0   (B) $\dfrac{1}{2}$ (C) 1   (D) 2   (E) 3

5. Evaluate: $\displaystyle\lim_{x\to 0}\frac{1-\cos 2x}{x\sin x}$

(A) -2   (B) 1   (C) 0   (D) undefined   (E) 2

6. Evaluate: $\displaystyle\lim_{x\to 0}\frac{\sin 3x}{\sin 4x}$

(A) undefined   (B) $-\dfrac{7}{11}$ (C) $\dfrac{3}{4}$ (D) 0   (E) 1

7. Evaluate: $\displaystyle\lim_{x\to 0}\frac{\sin(\sin x)}{\sin x}$

(A) 0   (B) undefined   (C) $+\infty$ (D) 1   (E) $\dfrac{1}{2}$

8. The position function of a particle that moves in a straight line is

$x(t)=2\pi t+\cos(2\pi t)$.

a) Find the velocity of the particle at time t.

b) Find the acceleration of the particle at time t.

c) For what values of t in the interval $0\le t\le 3$ is the particle at rest?

d) What is the maximum velocity of the particle?

9. For $f(x)=\sin^2 x$ and $g(x)=0.5x^2$ on the interval $\left[-\dfrac{\pi}{2},\dfrac{\pi}{2}\right]$, the

instantaneous change of rate of f is greater than the instantaneous change of

rate of g for which value of x?

(A) -0.8   (B) 0   (C) 0.9   (D) 1.2   (E) 1.5

10. Find the following limit (if it exists) without the aid of a calculator:

$\displaystyle\lim_{x\to\frac{\pi}{2}}\left(x-\frac{\pi}{2}\right)\tan x$

* 11. A particle moves along the x-axis so that at time t its position function is

$$x(t) = \sin\left(\pi t^2\right) \text{ for } 0 \le t \le 2.$$

a) Find the velocity of the particle at time t.

b) Find the acceleration of the particle at time t.

c) For what values of t does the particle change direction?

d) Find all values of t for which the particle is moving to the left.

Answers to Worksheet 3

1. D

2. B

3. B

4. B

5. E

6. C

7. D

8. a) $2\pi\left(1 - \sin\left(2\pi t\right)\right)$

b) $-4\pi^2 \cos\left(2\pi t\right)$

c) $t = \dfrac{1}{4}$ or $\dfrac{5}{4}$ or $\dfrac{9}{4}$

d) 4π

9. C

10. -1

11. a) $2\pi t \cos\left(\pi t^2\right)$

b) $2\pi\left(-2\pi t^2\left(\sin\left(\pi t^2\right)\right) + \cos\left(\pi t^2\right)\right)$

c) $t = \dfrac{1}{\sqrt{2}}, \dfrac{\sqrt{3}}{\sqrt{2}}, \dfrac{\sqrt{5}}{\sqrt{2}}, \dfrac{\sqrt{7}}{\sqrt{2}}$

d) $\dfrac{1}{\sqrt{2}} < t < \dfrac{\sqrt{3}}{\sqrt{2}}$ or $\dfrac{\sqrt{5}}{\sqrt{2}} < t < \dfrac{\sqrt{7}}{\sqrt{2}}$

Worksheet 4

1. Suppose that a person's blood pressure at time t (in seconds) is

    $$p = 100 + 18\sin 7t$$

    a) Find the maximum value of p (systolic pressure) and the minimum value

    of p (diastolic pressure).

    b) How many heartbeats per minute are predicted by the formula for p?

2. A drawbridge with arms of length 10 m is constructed as shown in the

    diagram. If the arms rotate at $\dfrac{\pi}{60}$ radians/s as the bridge is raised, how fast is

    the distance AB changing when the angle of inclination is $\theta = \dfrac{\pi}{6}$ radians?

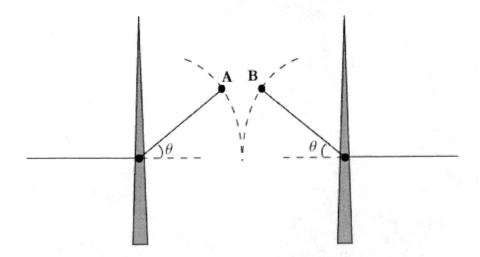

3. Find all local extrema of the function

    $$f(x) = \sin x(1 - \cos x), \text{ where } 0 \le x \le 2\pi$$

4. The height of an object thrown downward from an initial altitude of 200 m is

$$h(t) = 200 - 12t - 5t^2$$

The object is being tracked by a searchlight 100 m from where the object will hit the ground. How fast is the angle of elevation of the searchlight changing after 3 s?

5. Find the maximum value of $3\sin x + 4\cos x$.

6. Find the maximum value of $4\sin x + 3\cos x$.

7. The hypotenuse of a right angled triangle is 12 metres. Find the angles of the triangle so that a) the area and b) the perimeter are maximized. Find the maximum area and maximum perimeter.

8. A triangle has 2 fixed equal sides. Find the angle between those sides so that the area of the triangle is maximized.

9. A kite 120 feet above the ground is moving horizontally at the rate of 10 feet per second. At what rate is the angle of the string to the horizontal diminishing when the angle is 30 degrees?

10. A rotating beacon of light is situated 3600 feet off a straight shore. If the light from the beacon turns at 4π radians per minute, how fast does the beam of light sweep along the shore at a) its nearest point and b) at a point 4800 feet from the nearest point?

11. Use trigonometry to find the maximum area of a trapezoid three of whose edges are 10 metres long.

12. AB is a variable line passing through (16,2) where A and B are the positive x and y intercepts of the line. Find the minimum length of AB.

13. a) Find the smallest positive x value at which the graphs of $y = \operatorname{cosec} x$ and $y = 2\cos x$ intersect.

   b) Find the slopes of each at the point of intersection.

   c) Comment on what you found in part b).

14. The graph of the function $f(x)$ defined below is continuous and differentiable everywhere. Find the value of the constants a and b.

$$f(x)\begin{cases} ax^2 + bx & \text{for } x \leq 1 \\ \cos(\pi x) & \text{for } x > 1 \end{cases}$$

15. Find the angle of intersection of the graphs of $y = 2\cos x$ and $y = \sec x$ in

the interval $0 \le x \le \dfrac{\pi}{2}$.

16. In a 100 m race, the position of the Canadian runner after t seconds is given

by the formula $C(t) = 100 - 100\cos\left(\dfrac{\pi t}{20}\right)$ where the position is in metres.

During the same race, the position of the American runner is given by the

formula $A(t) = 50 - 50\cos\left(\dfrac{\pi t}{10}\right)$.

a) Graph the two equations on your calculator for $0 \le t \le 10$ and explain in

words what happened during the race.

b) Find the maximum lead either runner had during the race. At what time

did this take place?

c) At what instant in the race is the velocity of the Canadian runner equal to

her average velocity for the entire race?

Answers to Worksheet 4

1. a) 118 (max) 82 (min)

   b) 67 (approx.)

2. $\dfrac{\pi}{6}$ m/sec

3. $\left(\dfrac{2\pi}{3}, \dfrac{3\sqrt{3}}{4}\right)$ maximum, $\left(\dfrac{4\pi}{3}, -\dfrac{3\sqrt{3}}{4}\right)$ minimum

4. 0.174 radians/sec

5. 5

6. 5

7. a) 45°, 45°, 90°   b) 45°, 45°, 90°

   maximum area = 36   maximum perimeter $= 12\left(\sqrt{2}+1\right)$

8. 90°

9. $\dfrac{1}{48}$

10. a) $14400\,\pi$,   b) $40000\,\pi$

11. $75\sqrt{3}$

12. $10\sqrt{5}$

13. a) $\dfrac{\pi}{4}$ b) $-\sqrt{2}$ and $-\sqrt{2}$ c) They are tangent to one another.

14. $a=1$, $b=-2$

15. 70.5°

16. a) At first, the American runner runs faster than the Canadian. However, in the end, the Canadian runner catches up and ties the American runner. They each complete the race in 10 seconds.

    b) The maximum lead is 25 metres at approximately 6.67 seconds.

    c) 4.39 seconds

Worksheet 5

1. The minute hand of a clock is 6 cm long and the hour hand is 4 cm long. If the hands move continuously,

    a) at what rate in radians per hour does the angle between the hands change?

    b) at what rate in cm per hour is the distance between the tips of the hands changing at 2 o'clock?

2. The minute hand of a clock is 4 inches and the hour hand is 3 inches long. How fast are the tips of the hands separating at 9 p.m.?

3. An isosceles triangle is inscribed inside a circle of radius 2 cm. Find the maximum area of the triangle.

4. Find a relative minimum point on a graph of $y = \dfrac{1}{2}\cos 2x + \cos x$.

5. A lighthouse has a revolving light which turns at the rate of 2 revolutions per minute. The lighthouse is situated $\dfrac{1}{2}$ mile from a straight beach. Find how fast the spot of light from the beam is moving along the beach when it is 1 mile from the point of the beach nearest the lighthouse.

6. Use trigonometry to find the maximum lateral surface area of a cylinder inscribed inside a sphere of radius $4\sqrt{3}$ cm. Find also the maximum volume of a cylinder inscribed in the same sphere.

7. Find the maximum volume of a cone inscribed in a sphere of radius 9 cm.

8. Triangle *ABC* is isosceles and has its base on the extension of a diameter of a semi-circle. The radius of the semi-circle is 6 cm. Find the minimum area of triangle *ABC*.

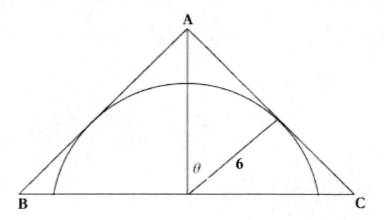

9. Find the shortest possible ladder which can reach a house if there is a wall 8 feet high, which is located 1 foot away from the house.

10. A ladder 27 feet long is leaning on and over a fence. If the ladder remains in contact with the top of the fence which is 8 feet high and the lower end of the ladder is pulled directly away from the fence, what is the greatest <u>horizontal</u> distance the ladder can project over the fence?

11. A sign board 45 feet high stands at the top of a cliff 86 feet high. How far from the foot of a cliff should a man stand in order for the sign to subtend the largest possible angle at his eyes which are placed 6 feet above the ground?

12. Given the circle $x^2 + y^2 = r^2$ and the point A ($a,0$) where $a < r$;

Let P be an arbitrary point on the circle. Prove that the <u>maximum</u> angle *OPA* occurs when angle $OAP = 90°$.

13. AB is a ladder 20 feet long which is pushed down the wall so that the angle θ

between the ladder and the wall increases uniformly at $\dfrac{\pi}{12}$ radians per second.

Find the velocity and acceleration of A when AB is vertical. Point A is at the

bottom of the ladder.

14. A sphere is inscribed in a cone. If the total surface area of the cone is a

minimum, show that the semi-vertical angle α of the cone is $\text{Arcsin}\dfrac{1}{3}$.

Answers to Worksheet 5

1. a) $\dfrac{11\pi}{6}$ radians/hour   b) 22.6

2. $\dfrac{22\pi}{5}$ inches/hour

3. $3\sqrt{3}$ 8. 72

4. $\left(\dfrac{2\pi}{3}, -\dfrac{3}{4}\right)$ 9. $5\sqrt{5}$

5. 10π miles/minute   10. $5\sqrt{5}$

6. a) 96π b) 256π 11. 100 ft

7. 288π 13. velocity is $\dfrac{5\pi}{3}$ ft/sec, acceleration is 0

Worksheet 6

1. Show that $x = a\cos\theta$, $y = b\sin\theta$ is a parametrization of the ellipse

   $\dfrac{x^2}{a^2} + \dfrac{y^2}{b^2} = 1$. Use the parametrization to find the shape and area of the largest

   rectangle that can be inscribed inside the ellipse.

2. If $y = \dfrac{\sin x}{x}$ show that $x\dfrac{d^2 y}{dy^2} + 2\dfrac{dy}{dx} + xy = 0$.

3. In the diagram below the centre of the flywheel is fixed and the cylinder is

   fixed. The piston is moving directly towards the centre of the wheel at the

   rate of 50 cm per second. The connecting rod, AB, is 70 cm long and the

   radius of the wheel is 30 cm. How fast is the wheel rotating when $\theta = \dfrac{\pi}{3}$

   radians, i.e. $60°$?

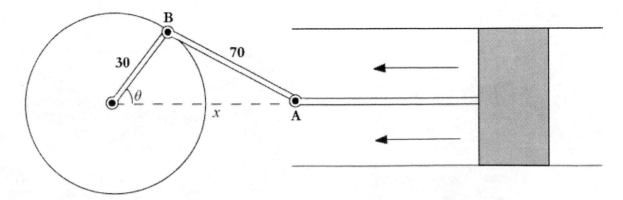

4. Given a circle of radius r and a tangent line ℓ to the circle through a given point P on the circle. From a variable point R on the circle a perpendicular RQ is drawn to ℓ with Q on ℓ. Determine the maximum area of $\triangle PQR$.

5. In a circular arena there is a light at L. A boy starting at P runs at a rate of 4 metres per second towards the centre O. At what rate will his shadow be moving along the side of the arena when he is halfway from P to O? (see diagram)

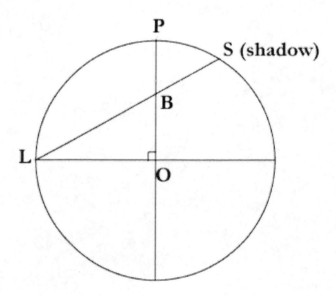

6. A runner is running around a circular track of radius 50 metres at a speed of 5 m/s sec. His father is watching from a point 30 metres outside the track. How quickly is the distance between father and son changing at the instant when the son is 70 m from his father?

7. (calculator needed) a) Find the point on the curve $y = \sin x$ nearest to (0,1).

   b) Find the minimum distance from (0,1) to the curve $y = \sin x$.

8. A circular table has radius 1 metre. A light L is suspended above the centre of the table such that the illumination at a point P on the perimeter of the table is proportional to $\cos\theta$ and inversely proportional to $\sqrt{\ell}$. Find the height h to maximize the illumination on the perimeter of the table.

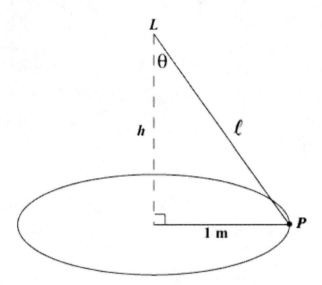

Answers to Worksheet 7

1. slope is $\sqrt{2}a$ by $\sqrt{2}b$ for a maximum area of $2ab$

3. 1.56 radians/second

4. $0.6495r^2$ or $\dfrac{3\sqrt{3}}{8}r^2$ (exact answer)

5. 6.4 m/sec

6. $\dfrac{20\sqrt{3}}{7}$

7. a) (0.4787, 0.4606) b) 0.721

8. $\sqrt{2}$ m

CHAPTER 10

Logarithms

It is assumed that students are familiar with the basic concepts and formulae associated with logarithms.

For example $\log_b n = x$ means $b^x = n$

and that $10^{\log x} = x$.

Also $\log a + \log b = \log(ab)$

$$\log a - \log b = \log\left(\frac{a}{b}\right)$$

$$\log a^n = n \log a$$

Also $\log_b a = \dfrac{\log_c a}{\log_c b}$ where c is any positive number

$y = \log x$ is the inverse of the function $y = 10^x$

We will start by finding the slope of $y = a^x$. Note that we cannot use the formula $D(x^n) = nx^{n-1}$ since this formula only applies when n is a constant.

<u>To Investigate the Derivative of $f(x) = a^x$</u>

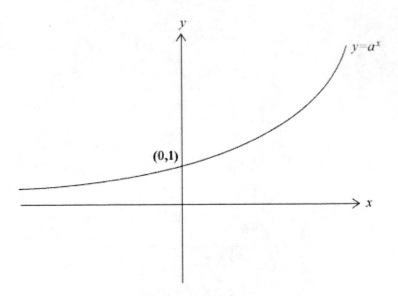

The definition of $f'(x)$ is $\lim\limits_{h \to 0} \dfrac{f(x+h) - f(x)}{h}$.

$$\therefore \quad D\left(a^x\right) = \lim_{h \to 0} \frac{a^{x+h} - a^x}{h}$$

$$= \lim_{h \to 0} \frac{a^x \cdot a^h - a^x}{h}$$

$$= \lim_{h \to 0} \frac{a^x \left(a^h - a^0\right)}{h}$$

But a^x does not vary with variation in the value of h

$$\therefore \quad \text{We can write: } D\left(a^x\right) = a^x \lim_{h \to 0} \frac{\left(a^h - a^0\right)}{h}$$

By definition, note that $\lim\limits_{h \to 0} \dfrac{\left(a^h - a^0\right)}{h}$ is the slope of $y = a^x$ at $(0,1)$.

Call this value L.

$$\therefore \quad D\left(a^x\right) = a^x \cdot L$$

✳

This means for all positive values of a, the slope of $y = a^x$ at any point is a^x times

(the slope of the graph at (0,1)). Let's see what happens to the slopes of $y = a^x$ at

(0,1) for different positive values of a as suggested by the graphs shown below.

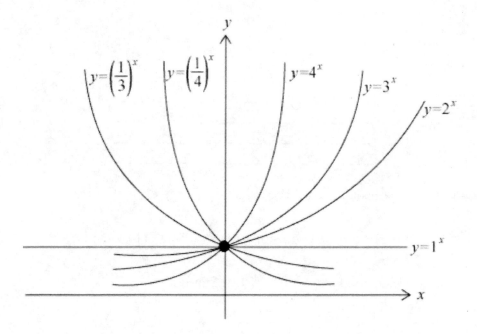

The slope of $y = \left(\dfrac{1}{3}\right)^x$ at (0,1) is clearly negative.

The slope of $y = 1^x$ at (0,1) is clearly 0.

The slope of $y = 2^x$ at (0,1) is clearly positive.

The slope of $y = 3^x$ at (0,1) is clearly a larger positive.

Let's look at a table of values of slopes of $y = a^x$ at $(0,1)$ for different values of a.

a	$f'(0)$
$\frac{1}{4}$	-1.386
$\frac{1}{3}$	-1.099
$\frac{1}{2}$	-0.693
1	0
2	+0.693
3	+1.099
4	+1.386

From the tables of values and from the graphs it seems clear that there exists a value

of a (between 2 and 3) such that the slope of $f(x) = a^x$ at $(0,1)$ is 1.

We will look at a table of values of a between 2 and 3.

a	$f'(0)$
2.5	0.916
2.6	0.956
2.7	0.993
2.8	1.030

When a is between 2.7 and 2.8 the slope of $y = a^x$ at $(0,1)$ is 1. We call this special

value e. This means that $\boxed{D(e^x) = e^x}$ from ✱ earlier.

Now let's consider the function $y = \log_e x$ and its derivative. $\log_e x$ is conventionally

written $\ln x$.

Let $y = \ln x$.

Then $e^y = x$.

Differentiate with respect to x.

$$e^y \left(\frac{dy}{dx} \right) = 1$$

$$\therefore \quad \frac{dy}{dx} = \frac{1}{e^y} = \frac{1}{x}$$

$$\therefore \quad \boxed{D(\ln x) = \frac{1}{x}}$$

Now let's return to $D\left(a^x\right)$.

Let $\quad y = a^x$

Then $\quad \ln y = \ln\left(a^x\right)$

i.e. $\quad \ln y = x \ln a$

Differentiate with respect to x.

$$\frac{1}{y}\frac{dy}{dx} = \ln a \quad \text{(remember that } \ln a \text{ is a constant)}$$

$$\therefore \quad \frac{dy}{dx} = y \ln a = a^x \ln a$$

$$\therefore \quad \boxed{D\left(a^x\right) = a^x \ln a}$$

Note incidentally that this establishes in fact that the slope of $y = a^x$ at $(0,1)$ is $\ln a$.

Example

Find $\quad D\left(e^{2x}\right)$.

$$D\left(e^{2x}\right) = e^{2x} \cdot 2 \quad \text{by Chain Rule}$$

Example

Solve $e^{2x} = 5$.

Take logs (base e) of both sides then

$$\ln\left(e^{2x}\right) = \ln 5$$

$\therefore \qquad 2x \ln e = \ln 5$

$\therefore \qquad 2x = \ln 5$

$$x = \frac{\ln 5}{2} = 0.8047 \quad \text{(approx.)}$$

Example

Find $\lim\limits_{x \to 0} \dfrac{2^x - 1}{3^x - 1}$.

By L'Hôpital's Rule the limit $= \lim\limits_{x \to 0} \dfrac{2^x \ln 2}{3^x \ln 3} = \dfrac{\ln 2}{\ln 3} = 0.631 \quad \text{(approx.)}$

Example

Find an equation for the tangent to $e^{\sin x} \cdot y + \ln y - y = 1$ at $(0, e)$.

Differentiate with respect to x:

$$y e^{\sin x} \cos x + e^{\sin x} \frac{dy}{dx} + \frac{1}{y} \frac{dy}{dx} - \frac{dy}{dx} = 0$$

At $(0, e)$ we have $e \cdot e^{\sin 0} \cdot \cos 0 + e^{\sin 0} \dfrac{dy}{dx} + \dfrac{1}{e} \dfrac{dy}{dx} - \dfrac{dy}{dx} = 0$

i.e. $\qquad e + \dfrac{dy}{dx} + \dfrac{1}{e} \dfrac{dy}{dx} - \dfrac{dy}{dx} = 0$

i.e. $\qquad \dfrac{dy}{dx} = -e^2$

Equation of tangent is $y - e = -e^2 (x - 0)$

i.e. $\qquad y = -e^2 x + e$

Worksheet 1

1. a) If $x = 2\ln 2$ what is the value of e^x?

   b) If $x = \dfrac{1}{2}\ln 64$ what is the value of e^{3x}?

2. Find the angle at which $y = \ln x$ intersects the x-axis.

3. Find $\dfrac{dy}{dx}$ if $y =$

   a) $\ln(\tan x)$ b) $\dfrac{x}{\ln x}$ c) $e^{\ln x}$ d) $e^{\ln x^2}$ e) $\left(e^{\ln x}\right)^2$

   f) e^2 g) $e^{\frac{1}{x}}$ h) $\ln 2x$ i) $\ln 3x$ j) $\ln\operatorname{cosec}\left(2x^3\right)$

4. Find $f'(x)$ if $f(x) =$

   a) $e^{\sin 2x}$ b) $\ln e^{-x}$ c) $\log_e x$ d) $\log_2 x^2$ e) $e^{-\ln x}$

5. a) Does $e^x = -3$ have a solution for x?

   b) Does $\ln(-3)$ exist as a real number?

   c) For what values of x is $\ln x$ defined?

   d) For what values of x does e^x make sense?

6. Simplify:

   a) $e^{\ln x}$ b) $(\ln e)^x$ c) $\ln\left(e^x\right)$ d) $\ln 1$

   e) $e^{2\ln 2}$ f) $e^{3\ln 3}$ g) $e^{\ln 10}$

7. If $f(x) = \ln x$ and $g(x) = e^x$ what is $f(g(x))$, $g(f(x))$?

8. Find: a) $\lim\limits_{x\to\infty} xe^{-x}$ b) $\lim\limits_{x\to 0} x{\cdot}e^{-x}$ c) $\lim\limits_{x\to\infty}\left(e^{\frac{1}{x}}-1\right)x$

9. a) Solve $e^x = 12$ using a calculator.

   b) Solve $e^{2x} = 2$ using a calculator.

10. Solve $\ln x = -2$ using a calculator.

11. a) Solve $e^x + e^{2x} = 2$.

    b) Solve $\log_2 x + \log_3 x = \dfrac{1}{\ln 2} + \dfrac{1}{\ln 3}$.

12. Simplify: a) $\dfrac{\log_{10} x}{\log_{10} e}$ b) $\dfrac{e^{2\ln 3} + \ln 1}{\ln\sqrt{e} + e^0}$ c) $\dfrac{\ln e^x + x\ln e}{\ln e^2}$

13. Find:

   a) $D\left(e^{2x}\right)$ b) $D\left(2^{2x}\right)$ c) $D\left(\ln x^x\right)$ d) $D\left(\ln\sec x\right)$

   e) $D\left(\sin(\ln x)\right)$ f) $D\left(\log_e x\right)$ g) $D\left(\log e^x\right)$ h) $D\left(e^{2x^2}\right)$

   i) $D\left(7^{3x}\right)$ j) $D\left(\cos\left(e^x\right)\right)$ k) $D\left(x^2{\cdot}2^x\right)$

   l) $D\left(\ln(x^2){\cdot}(2^x)\right)$ m) $D\left(\log e\right)$ n) $D\left(\ln nx\right)$, where n is a constant

Answers to Worksheet 1

1. a) 4   b) 512

2. $45°$

3. a) $\dfrac{\sec^2 x}{\tan x}$ b) $\dfrac{\ln x - 1}{(\ln x)^2}$ c) 1   d) $2x$ e) $2x$

    f) 0   g) $-\dfrac{e^{\frac{1}{x}}}{x^2}$ h) $\dfrac{1}{x}$ i) $\dfrac{1}{x}$ j) $-6x^2 \cot(2x^3)$

4. a) $e^{\sin 2x} \cdot 2\cos 2x$ b) -1   c) $\dfrac{1}{x}$ d) $\dfrac{2}{x \ln 2}$ e) $-\dfrac{1}{x^2}$

5. a) No   b) No   c) $x > 0$ d) all values of x

6. a) x b) 1   c) x d) 0

    e) 4   f) 27   g) 10

7. x, x

8. a) 0   b) 0   c) 1

9. a) 2.485   b) 0.34657

10. 0.1353

11. a) $x = 0$ only   b) $x = e$

12. a) $\ln x$ b) 6   c) x

13. a) $2e^{2x}$ b) $2^{2x+1} \ln 2$ c) $\ln x + 1$ d) $\tan x$

    e) $\dfrac{\cos(\ln x)}{x}$ f) $\dfrac{1}{x}$ g) $\log e$ h) $4x \cdot e^{2x^2}$

    i) $3 \cdot 7^{3x} \ln 7$ j) $-e^x \cdot \sin e^x$ k) $2^{x+1} \cdot x + x^2 \cdot 2^x \ln 2$

    l) $2^{x+1} \cdot \dfrac{1}{x} + 2^{x+1} \ln x \cdot \ln 2$ m) zero   n) $\dfrac{1}{x}$

Differentiation of the Inverse of a Function

Remember that $f(x) = e^x$ and $g(x) = \ln x$ are inverse functions of each other.

We will now look at the derivative of an inverse function in general and see how it relates to the exponential and logarithmic function.

If $g(x)$ and $f(x)$ are inverses of each other then $g'(a) = \dfrac{1}{f'(b)}$ **where** $g(a) = b$.

In effect in the diagram below we are saying that the slope of the tangent at point A is the reciprocal of the slope of the tangent at B (and vice versa).

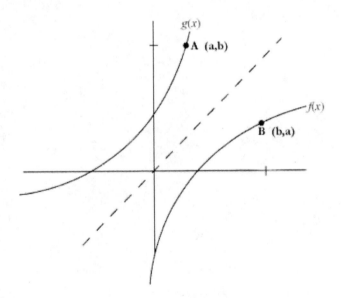

Proof: Let f and g be inverses. Thus $f(g(x)) = x$.

Differentiate with respect to x then

$$f'(g(x))g'(x) = 1 \quad \text{Chain Rule}$$

$$\therefore \quad f'(g(x)) = \frac{1}{g'(x)}$$

Let $g(a) = b$ then $f'(b) = \dfrac{1}{g'(a)}$ \quad i.e. \quad $g'(a) = \dfrac{1}{f'(b)}$.

The slope of an inverse function at (say) (2,3) is the reciprocal of the slope of the original function at (3,2).

Example

Question: Let $f(x)=\ln(x+1)$ then the inverse of f is $g(x)=f^{-1}(x)=e^x-1$.

Note that, for example, $g(2)=e^2-1$.

$$f'(x)=\frac{1}{x+1} \text{ and } g'(x)=e^x$$

i.e. $f'(e^2-1)=\dfrac{1}{e^2-1+1}$ and $g'(2)=e^2$

$$=\frac{1}{e^2}$$

Note that $f'(e^2-1)$ is the reciprocal of $g'(2)$

Example

Question: Let $f'(x)=x^3+x$, find $g'(2)$ where $g(x)=f^{-1}(x)$.

Answer: To find $g(x)$ in explicit terms of x is not practical and hence we will use the preceding theorem. Note that $f'(x)=3x^2+1$ is always positive and hence f is a monotonic, increasing function and hence f^{-1} exists.

Note that $f(1)=2$ and hence $g(2)=1$.

It follows that $g'(2)=\dfrac{1}{f'(1)}=\dfrac{1}{3(1)^2+1}=\dfrac{1}{4}$. It should always be remembered of course that this method only applies where f has an inverse which exists as a function.

Example

Question: Find the slope of $2^x + 3^y = 2$ at the point $(0,0)$.

Answer: Differentiate with respect to x then

$$2^x \ln 2 + 3^y \ln 3 \frac{dy}{dx} = 0$$

At $(0,0)$ $\ln 2 + \ln 3 \frac{dy}{dx} = 0$

$$\therefore \qquad \frac{dy}{dx} = -\frac{\ln 2}{\ln 3} = -0.631 \text{ (approx.)}$$

Example

Question: Find the value of $\lim\limits_{x \to 0} \dfrac{e^{\tan x} - 1}{\tan x}$

Answer: As $x \to 0$, the value of the limit "becomes" $\dfrac{0}{0}$ and hence we can use

L'Hôpital's Rule

i.e. $\text{limit} = \lim\limits_{x \to 0} \dfrac{e^{\tan x} \sec^2 x}{\sec^2 x} = \lim\limits_{x \to 0} e^{\tan x} = e^0 = 1$

Example

Question: Find where $y = 2e^{-x}$ intersects $y = e^x - 1$.

Answer: The intersection point(s) occur(s) when

$$2e^{-x} = e^x - 1$$

i.e. $0 = e^x - 1 - 2e^{-x}$

i.e. $\quad 0 = e^x - 1 - \dfrac{2}{e^x}$

$$0 = e^{2x} - e^x - 2$$

$$0 = \left(e^x + 1\right)\left(e^x - 2\right)$$

$\therefore \quad e^x = -1 \text{ or } e^x = 2$

But $e^x = -1$ does not have a solution and hence $e^x = 2$ is the only

solution i.e. $x = \ln 2$.

$\therefore \qquad$ The point of intersection is $\left(\ln 2, 1\right)$. Check this by using your

graphing calculator.

Example

Question: $\quad$ Graph $y = xe^{-x} = \dfrac{x}{e^x} = f(x)$

Answer: $\quad$ (Use your graphing calculator to check this work)

The most important characteristics of a graph as seen in Chapter 3 are

the intercepts, asymptotes (if any) and relative max/min points.

$$\text{For } y = \dfrac{x}{e^x} = f(x).$$

Intercepts	Asymptotes
$(0,0)$ only	As $x \to \infty$, $y \to 0$
	i.e. $y = 0$ is an asymptote.
	No other asymptotes exist
	since e^x is never zero.

To find relative max/min points

$$f'(x) = \frac{dy}{dx} = \frac{e^x \bullet 1 - x \bullet e^x}{e^{2x}} = \frac{1-x}{e^x}$$

$\frac{dy}{dx} = 0$ occurs when $x = 1$

$\therefore \quad \left(1, \frac{1}{e}\right)$ is a critical point.

Using a signed number line we have

x		1	
$f'(x)$	**positive**	0	**negative**

$\therefore \quad x = 1$ yields a relative maximum point.

Alternatively note that $f''(x) = \frac{x-2}{e^x}$ which is negative when $x = 1$,

i.e. $f''(1) = -\frac{1}{e}$ further justifying the fact that $x = 1$ yields a relative

maximum point. The graph of $y = \frac{x}{e^x}$ looks like:

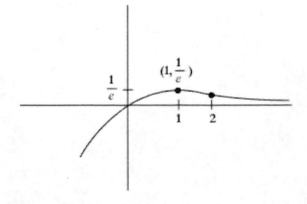

Example

Question: Graph $y = 2^{\frac{1}{x}}$ (check details using your graphing calculator)

Answer:

Intercepts	Asymptotes

None.

As $x \rightarrow +\infty$, $y \rightarrow 1^+$

As $x \rightarrow -\infty$, $y \rightarrow 1^-$

$\therefore y = 1$ is an asymptote.

No other asymptotes occur.

<u>To find relative max/min points</u>

$$\frac{dy}{dx} = 2^{\frac{1}{x}} \ln 2 \left(-\frac{1}{x^2} \right) = -\frac{\ln 2 \cdot 2^{\frac{1}{x}}}{x^2}$$

which is never zero and hence there are no relative max/min points.

The interesting aspect of this graph is the fact that as $x \rightarrow 0^+$, $y = +\infty$

but   as $x \rightarrow 0^-$, $y \rightarrow 0$

The graph looks like

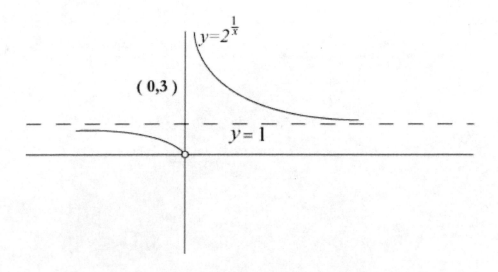

A Note of Interest:

<u>Graphing $y = \ln x$</u>

When considering parallel lines in early grades students are often told that two lines are parallel if they are the same distance apart. In co-ordinate geometry, they learn that parallel lines have equal slopes. As definitions in general they have their limitations when considering curved lines.

For example whether $x^2 + y^2 = 16$ and $x^2 + y^2 = 25$ are parallel is debatable using the "definitions" above. More instructive is the case of comparing

$$f(x) = \ln x \text{ and } g(x) = \ln 2x$$

$f'(x) = g'(x)$ for all x but few mathematicians would consider them parallel.

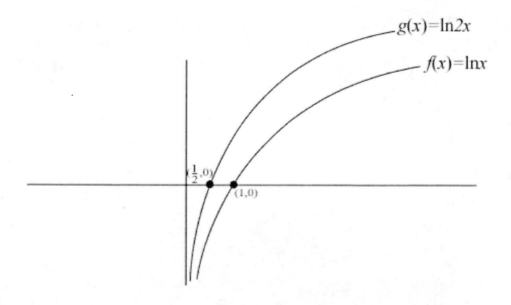

Worksheet 2 – Use algebraic techniques to find intercepts, asymptotes,

max/min values. You may check graphs using a graphing calculator.

1. Find the point of intersection of $y = e^x$ and $y = 2 + 3e^{-x}$. Find their angle of intersection.

2. Use logarithms to simplify finding $D\sqrt{\dfrac{x+1}{x-1}}$.

3. Find a) $D\left(\log 10^x\right)$ b) $D\left(10^x\right)$ c) $D\left(\log_e \dfrac{1}{x}\right)$

4. Find the slope of the tangent to $y = e^x$ at $(0,1)$.

5. a) Use logarithms to find $\dfrac{dy}{dx}$ if $y = x^x$.

    b) Find the minimum value of x^x.

    c) How many solutions are there to $x^x = 0.7$?

6. Graph $y = \dfrac{x}{e^x}$. Show maximum and minimum points (if any). Find the x and y intercepts. What happens to y as x approaches $\pm\infty$? Try to find an inflection point.

7. Graph $y = xe^x$.

8. Graph $y = x^2 e^{-x}$ showing max/min points, asymptotes (if any).

9. Graph $y = x^2 e^x$.

10. Find the maximum or minimum value of $2^x + 2^{-x}$. State which it is.

11. Sketch the graph of $y = xe^{-x^2}$.

12. Investigate whether $y = \dfrac{x}{\ln x}$ has a maximum or minimum point.

13. Find the minimum value of $e^x + e^{-x+1}$.

14. Graph $y = x \ln x$ for positive values of x. Show maximum or minimum points and intercept(s).

15. If $y = x^{\frac{1}{x}}$ find $\dfrac{dy}{dx}$ and hence deduce the minimum value of y (or is it the maximum value?)

16. Sketch $y = x^3 e^x$ locating maximum and minimum points but not inflection points.

17. Find x so that $\dfrac{\ln x}{x}$ has a maximum value. What is the maximum value?

18. Does $y = e^x \ln x$ have a maximum or minimum point?

19. Find a positive value of k so that $y = e^x$ intersects $y = k \sin x$ only once for $x > 0$.

Answers to Worksheet 2

1. $(\ln 3, 3)$, $63.4°$

2. $\sqrt{\dfrac{x+1}{x-1}} \cdot \dfrac{-1}{(x+1)(x-1)}$

3. a) $\log 10$ b) $10^x \ln 10$ c) $-\dfrac{1}{x}$

4. 1

5. a) $x^x (\ln x + 1)$ b) 0.6922 c) 2

6.

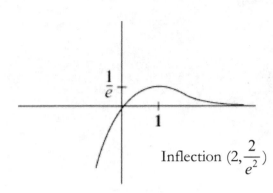

Inflection $(2, \frac{2}{e^2})$

7.

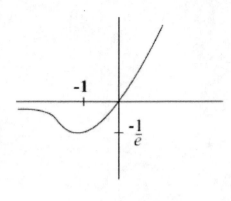

8.

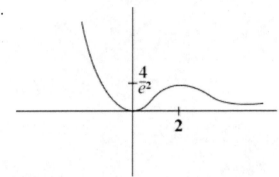

Inflection points occur when
$x = 2 + \sqrt{2}$ and $2 - \sqrt{2}$

9.

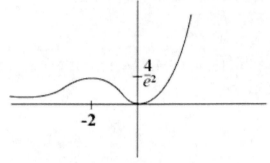

Inflection points occur when
$x = -2 + \sqrt{2}$ and $-2 - \sqrt{2}$

10. Minimum value of 2

11.

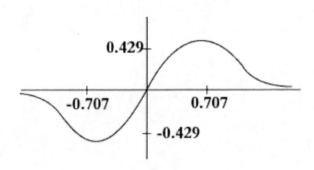

12. Minimum point at (e, e)

13. $2\sqrt{e}$ (i.e. 3.2974)

14. x intercept is $(1,0)$

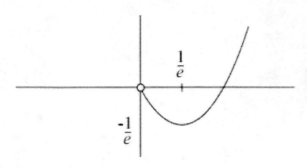

17. x is e. Maximum

18. No since $\ln x + \dfrac{1}{x}$ is never zero

19. $\sqrt{2}e^{\frac{\pi}{4}}$

15. Maximum value of $e^{\frac{1}{e}}$ (i.e. 1.4446679)

16.

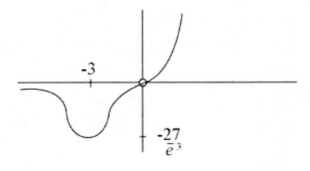

Worksheet 3

1. Let $f(x) = x^3 + 2x$ and let $g(x)$ be the inverse of $f(x)$. Find $g'(3)$.

2. Let $f(x) = x + 2^x$ and let $g(x)$ be the inverse of $f(x)$. Find $g'(3)$.

3. Show that if f is an increasing function then f^{-1} is also an increasing function.

4. Find the error in the following deductions:

   Let $f(x) = x^3 - x$ then $f(1) = 0$ and $f'(x) = 3x^2 - 1$

$$\therefore \quad \text{if } g \text{ is the inverse of } f \text{ then}$$

$$g'(0) = \frac{1}{f'(1)} = \frac{1}{3(1)^2 - 1} = \frac{1}{2}$$

5. Find $\dfrac{dy}{dx}$ if $y = \ln(\ln x)$.

6. Show that $y = \ln x - x + \dfrac{1}{2}(x-1)^2 + 1$ has a horizontal tangent and a point of

   inflection at (1,0).

7. If $y = \dfrac{e^{2x} - 1}{e^{2x} + 1}$ show that $\dfrac{dy}{dx} = 1 - y^2$.

8. a) Show that the graphs of $y = \ln(x^2 + 1)$ and $y = \dfrac{x^2}{e^x}$ are tangent to one

      another at $x = 0$.

   b) Are they internally or externally tangent?

Answers to Worksheet 3

1. $\dfrac{1}{5}$ 2. $\dfrac{1}{1 + \ln 4}$

3. If f in increasing, then $f'(x)$ in positive for all x, therefore $\dfrac{1}{f'(x)}$ is

   positive for all x $\therefore$ f^{-1} is increasing.

4. Error is that f does not have an inverse function.   5. $\dfrac{1}{x \ln x}$

8. b) Neither. They have a common tangent but cross at (0,0).

Some harder limits can be evaluated by the use of logarithms. The method tends to occur when we wish to find a limit of an expression to a power which is variable

e.g. $\lim\limits_{x \to 0}(1+x)^{\cot x}$

Assuming that $g(x)$ is continuous and takes on positive values then we note that

$$\lim_{x \to a}\left[\ln\left(g(x)\right)\right] = \ln\left(\lim_{x \to a} g(x)\right)$$

To find the value of $\lim\limits_{x \to 0}(1+x)^{\cot x}$

Let $\quad L = \lim\limits_{x \to 0}(1+x)^{\cot x}$

Then $\quad \ln L = \ln\left(\lim_{x \to 0}(1+x)^{\cot x}\right)$

$$= \lim_{x \to 0}\left(\ln(1+x)^{\cot x}\right)$$

$$= \lim_{x \to 0}\left(\cot x \ln(1+x)\right)$$

$$= \lim_{x \to 0}\left(\frac{\ln(1+x)}{\tan x}\right)$$

Using L'Hôpital's Rule

$$\ln L = \lim_{x \to 0}\frac{\dfrac{1}{1+x}}{\sec^2 x} = 1$$

$\therefore \qquad L = e$

$\therefore \qquad \lim\limits_{x \to 0}(1+x)^{\cot x} = e$

To Investigate the Numerical Value of e

Let e^x be represented by a polynomial function in x.

i.e. $\quad e^x = a_0 + a_1 x + a_2 x^2 + a_3 x^3 + ...$ $\quad$ ✳

But $\quad e^0 = 1$, $\therefore a_0 = 1$

Differentiating ✳ with respect to x yields

$$e^x = a_1 + 2a_2 x + 3a_3 x^2 + ...$$

As before $e^0 = 1$, $\therefore a_1 = 1$ also.

Differentiating again yields

$$e^x = 2a_2 + 6a_3 x + ...$$

And as before we can deduce $a_2 = \dfrac{1}{2}$, $a_3 = \dfrac{1}{6}....$, etc.

In general, $a_i = \dfrac{1}{i!}$

$$\therefore \quad e^x = 1 + x + \frac{x^2}{2!} + \frac{x^3}{3!} + \frac{x^4}{4!}....$$

This equation is true incidentally for <u>all</u> values of x.

Note that this expression for e^x can be written as

$$\boxed{e^x = \lim_{n \to \infty} \left(1 + \frac{x}{n} \right)^n}$$

which is deducible from the expansion of the right hand side using the Binomial

Theorem. This is left as an exercise for the reader.

In particular $e = e^1 = 1 + 1 + \dfrac{1^2}{2!} + \dfrac{1^3}{3!} + \dots$

≈ 2.718

e is a very important number in mathematics and has many applications. For example in compound interest we have the following:

If \$1 is invested at 100% for one year its value at the end of the year is \$2.

If the interest is compounded semi-annually then the \$1 becomes $\$1 \bullet \left(1 + \dfrac{1}{2}\right)^2 = \$2.25$.

Similarly if the interest is compounded monthly the \$1 becomes $\$1 \bullet \left(1 + \dfrac{1}{12}\right)^{12} = \$2.61$ (approx.)

If the interest is compounded daily then the \$1 becomes $\$1 \bullet \left(1 + \dfrac{1}{365}\right)^{365} = \$2.715$.

Compounding the interest continuously means \$1 becomes $\$ \lim\limits_{n \to \infty} \left(1 + \dfrac{1}{n}\right)^n$ which is $\$e$ from previous page. In fact $e \approx 2.718$ as before.

Worksheet 4 – No Calculators

1. Find $f'(x)$ if $f(x)=x^{e^x}$.

2. Find the following limits (if they exist)

   a) $\lim\limits_{x\to 0}\dfrac{1-e^{3x}}{\tan x}$

   b) $\lim\limits_{x\to 0}(1+2x)^{\frac{1}{x}}$

   c) $\lim\limits_{t\to 0}\left(e^t+te^t\right)^{\frac{1}{t}}$

   d) $\lim\limits_{x\to 0}\dfrac{e^x-e^{-x}}{3x}$

   e) $\lim\limits_{x\to 0}\dfrac{10^x-e^x}{x}$

   f) $\lim\limits_{x\to \pi}\dfrac{\ln(\cos 2x)}{(\pi-x)^2}$

   g) $\lim\limits_{x\to 0}(1+2x)^{\operatorname{cosec}x}$

   h) $\lim\limits_{x\to 0}(1+\sin x)^{\frac{1}{x}}$

   i) $\lim\limits_{x\to \infty}\left(1+\dfrac{2}{x}\right)^x$

   j) $\lim\limits_{x\to 0}(1+\sin 2x)^{\frac{1}{x}}$

   k) $\lim\limits_{x\to \infty}x\left(e^{\frac{1}{x}}-1\right)$

   l) $\lim\limits_{x\to \infty}x\left(2^{\frac{1}{x}}-1\right)$

3. Given $f(x)=3^{2x}$, find $f'(0)$.

   (A) $\ln 3$ (B) $\ln 9$ (C) $3\ln 3$ (D) $6\ln 3$ (E) $\ln 6$

4. Given $4+xy-e^{xy}=0$, then $\dfrac{dy}{dx}=$

   (A) $\dfrac{\ln y}{x}$ (B) $\dfrac{-y}{x}$ (C) $\dfrac{-x}{y}$ (D) $\ln xy$ (E) $\ln x+y$

5. Given $f(x)=(1+\ln x)^3$ where $x\ge 0$, find the point(s) of inflection on the

   graph of $f(x)$.

   (A) $(e,8)$ (B) $\left(\dfrac{1}{e},8\right)$ (C) $\left(\dfrac{1}{e},8\right)$

   (D) $(e,8),\left(\dfrac{1}{e},0\right)$ (E) $(e,8),\left(\dfrac{1}{e},8\right)$

6. Find the equation of a tangent line to $f(x) = xe^x$ when $x = 1$.

(A) $x - ey + e = 0$ (B) $x - \frac{1}{2}y + e = 0$ (C) $ex + y - e = 0$

(D) $ex - 2e^2y + 1 = 0$ (E) $2ex - y - e = 0$

7. Evaluate $\lim\limits_{x \to 0} \dfrac{1 - \cos 2x}{x^2}$

(A) 0   (B) 1   (C) $\dfrac{3}{2}$ (D) 2   (E) 4

8. If $y = 5^{(x^3 - 2)}$, then $\dfrac{dy}{dx} =$

(A) $(x^3 - 2)5^{(x^3 - 2)}$ (B) $3x^2 (\ln 5)5^{(x^3 - 2)}$ (C) $3x^2 5^{(x^3 - 2)}$

(D) $\ln 5 \cdot 5^{x^3 - 2}$ (E) $x^3 (\ln 5)5^{x^3 - 2}$

9. If $f(x) = e^{\sin x}$, how many zeros does $f'(x)$ have on the closed interval

$[0, 2\pi]$.

(A) 1   (B) 2   (C) 3   (D) 4   (E) 0

10. If $f(x) = \ln x - k\sqrt{x}$ has a local maximum at $x = 4$ then the value of k is

(A) -1   (B) $\dfrac{1}{2}$ (C) 1   (D) 4   (E) none of these

11. Which of the following functions grows faster than e^x as $x \to \infty$?

(A) x^4 (B) $\ln x$ (C) e^{-x} (D) 3^x (E) $\dfrac{1}{2}e^x$

12. Let f be defined such that $f(x) = \ln(3x+2)^k$ for some positive constant

   k. If $f'(2) = 3$ what is the value of k?

   (A) $\dfrac{\ln 3}{\ln 8}$ (B) $\ln 8$ (C) 4 (D) 8 (E) 16

13. $\displaystyle\lim_{x \to \infty} x^{\frac{1}{x}} =$

   (A) 0 (B) 1 (C) ∞ (D) e (E) none of these

14. $\displaystyle\lim_{x \to \infty} (\ln x)^{\frac{1}{x}} =$

   (A) 0 (B) 1 (C) e (D) ∞ (E) none of these

Answers to Worksheet 4

1. $f'(x) = x^{e^x}\left(\dfrac{e^x}{x} + e^x \ln x\right)$

2. a) -3   b) e^2 c) e^2 d) $\dfrac{2}{3}$

   e) $\ln 10 - 1$ f) -2   g) e^2 h) e

   i) e^2 j) e^2 k) 1   l) $\ln 2$

3. B   4. B   5. D   6. E   7. D

8. B   9. B   10. C   11. D   12. D

13. B   14. B

Worksheet 5 – No Calculators except *

1. An equation for the horizontal asymptote to the graph of $y = \dfrac{2 - e^{\frac{1}{x}}}{2 + e^{\frac{1}{x}}}$ is

    (A) $y = -1$ (B) $y = -\dfrac{1}{2}$ (C) $y = \dfrac{1}{3}$ (D) $y = \dfrac{1}{2}$ (E) $y = 1$

2. $f(x) = \begin{cases} 1 + e^{-x} & \text{if } 0 \le x \le 5 \\ 1 + e^{x-10} & \text{if } 5 < x \le 10 \end{cases}$

    Which of the following statements is (are) true?

    I. $f(x)$ is continuous in the interval $[0,10]$.

    II. $f'(x)$ is continuous in the interval $[0,10]$.

    III. The graph of $f(x)$ is concave up in the interval $[0,10]$.

    (A) I only   (B) II only   (C) I and III   (D) I, II, III   (E) II and III

3. $f(x) = \begin{cases} e^{-x} + 2 & \text{for } x < 0 \\ ax + b & \text{for } x \ge 0 \end{cases}$ and $f(x)$ is differentiable at $x = 0$ then $a + b =$

    (A) 0   (B) 1   (C) 2   (D) 3   (E) 4

4. The function $f(x) = \tan(3^x)$ has one zero in the interval $[0,1.4]$. The derivative at this point is

    (A) $\ln 3$ (B) π (C) $\pi \ln 3$ (D) 3^π (E) none of these

5. If $y = f(x) = (\ln x)^2$ then $f'(2) =$

(A) $-2\ln 2$ (B) $-\ln 2$ (C) $\ln 2$ (D) $2\ln 2$ (E) $\frac{1}{3}(\ln 2)^3$

6. Evaluate $\lim\limits_{x \to 0} \dfrac{e^x - \cos x}{x}$

(A) 0   (B) 1   (C) -1   (D) 2   (E) e

7. The slope of the tangent to the curve $y = (x)^{\sin x}$ at $x = \pi$ is:

(A) $-\ln \pi$ (B) $-\pi$ (C) -1 (D) 0   (E) e

8. The equation of the tangent to $\ln(x + y) - e^{-y} = -x$ at the point $(1, 0)$ is:

(A) $y = x$ (B) $y = 2x - 2$ (C) $y = x - 1$

(D) $y = -x + 1$ (E) $y = -2x + 2$

9. A population, in millions, at any time t is given by the formula

$P(t) = \dfrac{20(1 - e^{-2t})}{5 - e^{-t}}$. The $\lim\limits_{t \to \infty} P(t)$, in millions, is

(A) 20   (B) 4   (C) 5   (D) infinite   (E) 0

10. A function whose derivative is a constant multiple of itself is

(A) Periodic   (B) Linear   (C) Exponential   (D) Quadratic   (E) Logarithmic

11. If $f(x) = e^x \ln x$ then $f'(e) =$

(A) $e^{e-1} + e^e$

(B) $e^{e+1} + e^e$

(C) $e^e + e$

(D) $e^e + \dfrac{1}{e}$

(E) e^{e-1}

12. If $f(x) = \sqrt{e^{2x} + 1}$ then $f'(0) =$

(A) $\dfrac{\sqrt{2}}{4}$

(B) $\sqrt{2}$

(C) $\dfrac{\sqrt{2}}{2}$

(D) 1

(E) $-\dfrac{\sqrt{2}}{2}$

*13. A population grows according to the equation $P(t) = 6000 - 5500e^{-0.159t}$ for

$t \geq 0$, t measured in years. This population will approach a limiting value as time

goes on. During which year will the population reach <u>half</u> of this limiting value?

(A) Second   (B) Third   (C) Fourth   (D) Eighth   (E) Twenty-Ninth

14. For which pair of functions $f(x)$ and $g(x)$ respectively below is

$$\lim_{x \to \infty} \frac{f(x)}{g(x)} = 0?$$

(A) e^x , x^2

(B) e^x , $\ln x$

(C) $\ln x$, e^x

(D) x , $\ln x$

(E) 3^x , 2^x

*15. $\displaystyle\lim_{h \to 0} \frac{2^{3+h} - 8}{h} =$

(A) 0   (B) 1   (C) 5.5   (D) 5.74   (E) 7.25

16. $\lim\limits_{x\to 1}\dfrac{\ln x}{\sin(\pi x)} =$

    (A) 0 (B) 1   (C) e (D) π (E) $-\dfrac{1}{\pi}$

17. $\lim\limits_{x\to 0}(\sec x)^{\frac{1}{x^2}} =$

    (A) 0   (B) 1   (C) e (D) $\sqrt{e}$ (E) $-\dfrac{1}{2}$

18. $\lim\limits_{x\to 0}\left(\dfrac{1}{x}-\dfrac{1}{xe^{2x}}\right)$

    (A) 0   (B) 1   (C) e (D) 2   (E) $\dfrac{2}{3}$

Answers to Worksheet 5

1. C   10. C
2. C   11. A
3. C   12. C
4. C   13. C
5. C   14. C
6. B   15. C
7. D   16. E
8. D   17. D
9. B   18. D

Worksheet 6

1. Does $y = \dfrac{\ln x}{x^2}$ have a maximum value? If so, find it.

2. Does $y = \ln(\ln x)$ have a maximum or minimum point?

3. Find a positive value for x so that $y = x^2 10^{-x}$ has a relative maximum.

4. Graph $y = \dfrac{\ln x}{x}$ for positive values of x. Show intercept(s), relative

   maximum/minimum points, and asymptotes (if any).

5. Find the co-ordinates of a minimum point on the graph of $y = \sin(e^x)$. Is the

   function $f(x) = \sin(e^x)$ periodic? If so, what is the period?

6. Find: a) $D\left(\ln\sqrt{\dfrac{1}{x}}\right)$ b) $D\left(\ln(\sec x + \tan x)\right)$

7. Find a maximum or minimum point on the graph of $y = e^x + e^{-x}$.

   State which it is and graph the function.

8. Find $\dfrac{dy}{dx}$ if $y = (\sin x)^{\cos x}$.

9. Find $\dfrac{dy}{dx}$ if $\ln y = e^x$ a) in terms of x and y and b) in terms of x only.

10. Graph $y = x - \ln x$.

11. The famous bell curve with which all high school students are familiar is of

    the form $y = e^{\frac{-1}{2}x^2}$. Draw its graph.

12. The equation of the shape of a cable supporting a suspension bridge is of the

form $h = 100\left(e^{\frac{x}{2}} + e^{\frac{-x}{2}}\right)$. Find the minimum height h.

Answers to Worksheet 6

1. Maximum value is $\dfrac{1}{2e}$.

2. No

3. $x = \dfrac{2}{\ln 10}$

4.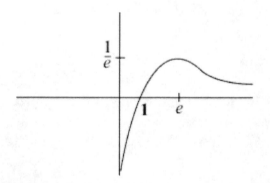

5. (1.55,-1) No

6. a) $-\dfrac{1}{2x}$ b) $\sec x$

7. Minimum point at (0,2)

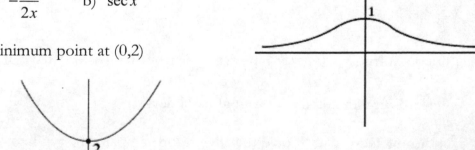

8. $(\sin x)^{\cos x} \cdot \left(\dfrac{\cos^2 x}{\sin x} - \sin x \cdot \ln(\sin x)\right)$

9. a) $y \cdot e^x$ b) $e^{e^x} \cdot e^x$ OR $e^{e^x + x}$

10.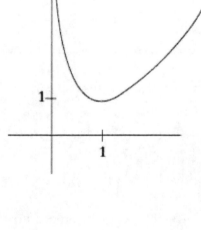

11.

12. 200

Worksheet 7

1. The Sundew Company estimates that the cost of producing x thousand cans

   of orange juice per month is given by the equation

   $C(x) = 1000x - 1000\ln(x-19) - 18000$. This equation is considered valid

   for $19.5 \le x \le 21$.

   a) Find the number of cans it should produce and the minimum monthly

      production cost.

   b) Find the minimum production cost per can of juice produced.

2. The percentage C of the population that has responded to a Cola

   advertisement after it has been marketed for t days is given by the formula

   $C = 0.7\left(1 - e^{-0.2t}\right)$. The marketing area contains 10 million potential customers

   and each response to the ad results in an average revenue to the company of

   \$ 0.70. The advertisement costs \$ 30000 to produce and \$ 500 per day to run.

   a) Find $\lim\limits_{t\to\infty} 0.7\left(1 - e^{-0.2t}\right)$ and write a one sentence interpretation of this result

   b) What percentage of potential customers has responded to the ad after 7

      days?

   c) Find a formula $P(t)$ that represents the **net profit** after t days of

      advertising. Find the net profit after 28 days.

   d) For how many days should the advertising campaign be run in order to

      maximize net profits? (No algebraic justification required, use a calculator)

*3. a) Find the minimum vertical distance between $y = e^x$ and $y = \ln x$.

    b) Find the minimum distance between $y = e^x$ and $y = \ln x$.

4. a) Find the maximum area of a rectangle whose four vertices lie on the

       x-axis, at (0,0), on the y-axis and on the graph of $y = e^{e-x}$.

    b) Find the minimum perimeter of a rectangle as defined in part a).

5. *a) Find the shortest distance from (0,0) to a point P on the curve $y = e^{-x^2}$.

    b) Using the point P found in part a), is OP a normal to the curve?

6. The equation $A(t) = \dfrac{0.22}{1 + 20e^{-0.9t}}$ models the amount of enzyme present in a

    chemical reaction after t hours $(t \geq 0)$.

    a) Find $\lim\limits_{t \to \infty} A(t)$.

    b) Find $\dfrac{dA}{dt}$.

    *c) Sketch a graph of $A(t)$.

    *d) Find the point P on the graph where A is increasing the fastest.

Answers to Worksheet 7

1. a) It should produce 20 000 cans. Minimum monthly cost is $2000.

    b) 10 ¢

2. a) 0.7%   b) 0.527%

    c) $P(t) = 49000\left(1 - e^{0.2t}\right) - 30000 - 500t$. The net profit after 28 days is $ 4818.

    d) 15 days

3. a) 2.330   b) $\sqrt{2}$ The points are (1,0) and (0,1).

4. a) e^{e-1}

    b) $2 + 2e$

5. a) 0.920   b) P is $\left(\sqrt{\ln\sqrt{2}}, \dfrac{1}{\sqrt{2}}\right)$. Yes OP is a normal.

6. a) 0.22   b) $\dfrac{3.96}{e^{0.9t}\left(1 + 20e^{-0.9t}\right)^2}$

    d) P is (3.33,0.11) (point of inflection)

A Note of Interest

e is a very important number in mathematics ranging from $D\!\left(e^x\right)=e^x$, to the

compound interest of 100% compounded continuously, to the idea as a corollary to

Moivre's Theorem that $e^{i\pi}=-1$ which combines together three fundamental

numbers of high school math.

Another example where e occurs, much less well known, is as follows:

We investigate the solutions to the equation

$\quad a^x = x^a$ where a is a positive number greater than 1.

For simplicity let's assume $x>0$.

For example $\quad x^2 = 2^x$ has $x=2$ and $x=4$ as solutions.

$\quad x^3 = 3^x$ has $x=3$ and $x=2.478$ as solutions.

$\quad x^4 = 4^x$ has $x=4$ and $x=2$ as solutions.

$\quad x^{100} = 100^x$ has $x=100$ and $x=1.0495$ as solutions.

It seems that there are always two positive solutions for x.

The question arises "Does there exist a value for a (greater than 1) so that the

solution to $a^x = x^a$ is unique, i.e. so that $y=a^x$ is tangent to $y=x^a$?"

We settle the matter with the following theorem:

Theorem

If x and a are real numbers greater than 1 then $x^a = a^x$ has two solutions except when $a = e$ when the solution is unique, namely, $a = e$.

"Proof"

Clearly $x^a = a^x$ has at least one solution i.e. $x = a$.

If $x^a = a^x$ has a unique solution then $y = x^a$ is tangent to $y = a^x$.

Therefore, at the point of tangency, the y values and the slopes are separately equal.

i.e. $\quad x^a = a^x \qquad$ (1)

and $\quad ax^{a-1} = a^x \ln a \qquad$ (2)

Substituting for a^x from (1) into (2) yields:

$$ax^{a-1} = x^a \ln a \qquad \textbf{✻}$$

Now since we know that $x = a$ is a solution then **if** the solution is unique then from ✻

$$a \cdot a^{a-1} = a^a \ln a$$

i.e. $\quad a^a = a^a \ln a$

Now a^a is not zero and hence $\ln a = 1$

$$\therefore \quad \underline{\underline{a = e}}$$

CHAPTER 11

Linear Tangent Approximations and Euler's Method

Before the arrival of calculators, a method for estimating values by extrapolation was

sometimes effected by the use of the fact that for small changes in x, $\dfrac{dy}{dx} \approx \dfrac{\Delta y}{\Delta x}$.

Graphically, this meant that on the graph below

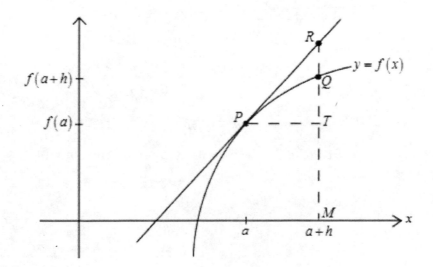

provided h was small, then points Q and R were virtually the same point. This

meant that their y co-ordinates were approximately equal.

i.e. $MQ \approx MR = (MT + TR)$

This means that $\boxed{f(a+h) \approx f(a) + f'(a)h}$

This approximation is called a linear approximation or a linear tangent

approximation. This is often written: $f(x + \Delta x) \approx f(x) + f'(x)\Delta x$.

Example

Question: Find, without the use of a calculator, an approximate value for $\sqrt{4.01}$.

Answer: Let $f(x) = \sqrt{x}$, let $a = 4$ and $h = 0.01$.

Note that $f'(x) = \dfrac{1}{2\sqrt{x}}$

Then $\sqrt{4.01} = f(4.01)$

$$= f(4 + 0.1)$$

$$\approx \sqrt{4} + \frac{1}{2\sqrt{4}}(0.01)$$

$$= 2 + 0.0025$$

$$= 2.0025$$

$\therefore$ $\sqrt{4.01} \approx 2.0025$ (This is an excellent approximation)

Example

Evaluate, approximately, the value of $x^{10} + 5x^5 + x$ when $x = 1.01$.

Let $f(x) = x^{10} + 5x^5 + x$.

Then $f'(x) = 10x^9 + 25x^4 + 1$

$$f'(1.01) \approx f(1) + f'(1)(0.01)$$

$$= 7 + 36(0.01)$$

$$\underline{\underline{= 7.36}}$$

In fact, $f(1.01) = 7.36967$ (approx.)

Example

The relation $x^2y + 2xy^3 = 8$ defines y as a function of x near to $(2,1)$.

Call this function $y = f(x)$. Use the linear tangent approximation to find an approximate value for $f(1.92)$.

$$f(1.92) \approx f(2) + f'(2)(-0.08) \quad \ast$$

To find $f'(2)$ we need to find $\dfrac{dy}{dx}$ when $x = 2$ and $y = 1$.

Differentiate with respect to x.

$$y2x + x^2\frac{dy}{dx} + 2\left(y^3 + x3y^2\frac{dy}{dx}\right) = 0$$

At $(2,1)$

$$4 + 4\frac{dy}{dx} + 2\left(1 + 6\frac{dy}{dx}\right) = 0$$

i.e. $\quad \dfrac{dy}{dx} = -\dfrac{3}{8}$

Substituting in $\ast$ yields:

$$f(1.92) \approx 1 + \left(-\frac{3}{8}\right)(-0.08)$$

$$\underline{\underline{= 1.03}}$$

Worksheet 1

1. Without the use of a calculator find the approximate value of

   a) $(8.02)^{\frac{1}{3}}$ b) $\sin 31°$ (note $1° = 0.01745$ radians)

   c) $(4.1)^{1.5}$ d) $\sqrt[3]{0.126}$

2. Find an approximate value for $x^3 - 3x^2 + 2x - 1$ when $x = 1.998$ without the aid of a calculator.

3. The surface area of a sphere is $4\pi r^2$. If the radius of the sphere is increased from 10 cm to 10.1 cm, what is the approximate increase in area?

4. One side of a rectangle is three times another side. If the perimeter increases by 2% what is the approximate percentage increase in area?

5. A new spherical ball bearing has a 3 cm radius. What is the approximate value of the metal lost when the radius wears down to 2.98 cm?

6. Find the percentage error in the volume of a cube if an error of 1% is made in measuring the edge of the cube.

7. (1,1) is a point on the graph of $x^2 y + y^2 x = 2$. Find a reasonable approximation for the y co-ordinate of a point near (1,1) whose x co-ordinate is 1.01.

8. The equation $x^4 + y + xy^4 = 1$ defines y implicitly in terms of x near the point (-1,1). Use the tangent line approximation at the point (-1,1) to estimate the value of y when $x = -0.9$.

9. The local linear approximation of a function f will always be greater than the function's value if, for all x in the interval containing the point of tangency,

(A) $f' < 0$ (B) $f' > 0$ (C) $f'' > 0$ (D) $f'' < 0$ (E) $f' = f'' = 0$

Answers to Worksheet 1

1. a) 2.0017   b) 5.151   c) 8.3   d) 0.5013

2. -1.004

3. 8π

4. 4.04%

5. 0.72π cubic cm

6. 3.03%

7. 0.99

8. 0.9

9. D

Euler's Method

Euler's Method involves the use of the linear tangent approximation more than once and is essentially the same. It is somewhat more accurate.

For example to find $\sqrt{4.02}$ where $f(x) = \sqrt{x}$ and $f'(x) = \dfrac{1}{2\sqrt{x}}$, $a = 4$ and $h = 0.02$,

we have $\qquad \sqrt{4.02} = f(4.02) \approx f(4) + f'(4)(0.02)$

$$= 2 + \frac{1}{2\sqrt{4}}(0.02) = 2.005$$

If we use Euler's Method where we do the approximation twice we have

$\underline{\Delta x = 0.01}$ repeated.

i.e. $\qquad \sqrt{4.01} = f(4.01) \approx f(4) + f'(4)(0.01)$

$$= 2 + 0.0025$$

Then $\quad \sqrt{4.02} = f(4.02) \approx f(4.01) + f'(4.01)(0.01)$

$$= 2.0025 + \frac{1}{2\sqrt{4.01}}(0.01)$$

$$= 2.0025 + \frac{1}{4.005}(0.01)$$

$$= 2.004997$$

In fact $\sqrt{4.02} = 2.0049938$ (approx.)

Example

Given $f'(x) = \dfrac{1}{x}$ and $f(1) = 0$ find $f(1.5)$ using Euler's Method with two

iterations of $\Delta x = 0.25$.

$$f(1.25) \approx f(1) + f'(1)(0.25)$$

$$= 0 + \frac{1}{1}(0.25)$$

$$= 0.25$$

$$f(1.5) = f(1.25) + \frac{1}{1.25}(0.25)$$

$$= 0.25 + \frac{4}{5}(0.25) = \underline{\underline{0.45}}$$

Graphically Euler's Method can be viewed and follows:

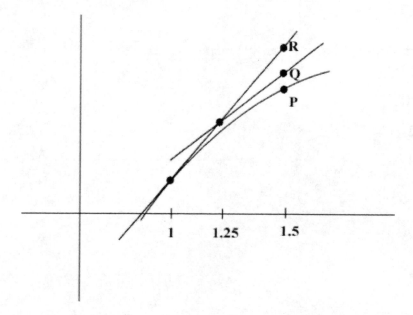

To evaluate $f(1.5)$: we want the y co-ordinate of P

Euler's method yields the y co-ordinate of Q

<u>One</u> linear approximation yields the y co-ordinate of R

Worksheet 2

1. $\dfrac{dy}{dx} = 0.5xy$. Use Euler's method to find y when $x = 2$ given that $y = 2$ when

   $x = 0$. Use steps each of size 1.

2. The solution of the differential equation $\dfrac{dy}{dx} = -\dfrac{x^2}{y}$ contains the point $(3,-2)$.

   Find using approximation methods the value of y when $x = 2.7$ with

   $\Delta x = -0.3$.

3. $\dfrac{dy}{dx} = \dfrac{x-y}{2y}$ and $y = -2$ when $x = 3$.

   An estimate for the value of y when $x = 3.2$ using a linear tangent

   approximation is:

   (A) -2   (B) -2.15   (C) -2.2   (D) -2.25   (E) -2.30

4.

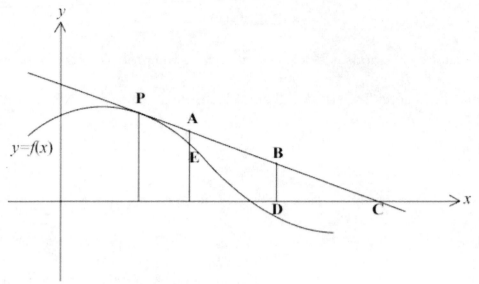

In the figure above PC is tangent to the graph of $y = f(x)$ at point P. Points

A and B are on PC and points D and E are on the graph of $y = f(x)$.

Which statement is true?

I.   Euler's method uses the y co-ordinate of point A to

approximate the y co-ordinate of point E.

II.   Euler's method uses the x co-ordinate of point B to

approximate the root of the function $f(x)$ at point D.

III.   Euler's method uses the x co-ordinate of C to approximate a

root of the function $f(x)$ at point D.

(A) I only   (B) II only   (C) III only   (D) I and II   (E) I and III

Answers to Worksheet 2

1. 3   2. -3.35   3. D   4. A

CHAPTER 12

Integration

For many practical situations it is necessary to reverse the process of differentiation.

For example, in Chapter 7 we learned that differentiating velocity with respect to time

produced acceleration. i.e. $a = \dfrac{dv}{dt}$.

However it is often necessary to "go backwards" from acceleration to velocity, since,

for example, we know that acceleration due to gravity (32 feet per second) per second

(approx) which is (9.81 metres per second) per second and it would be helpful to find

velocity and position after a certain period of time. Furthermore acceleration might

well not be constant and then finding velocity requires reversing the process of

differentiation.

Going backwards from differentiation is called INTEGRATION and is a major field of

study.

In this chapter we will concern ourselves merely with learning the algebraic

mechanics of integration rather than practical examples. We will study applications

of integration in later chapters.

If $$\frac{dy}{dx} = 2x$$

Then $y = x^2 + c$ where c is some constant

We write $\int 2x \, dx = x^2 + c$.

$\int f(x) \, dx$ tells us that we are finding out what expression it is that when

differentiated becomes $f(x)$.

e.g. $\int 3x^2 \, dx = x^3 + c$

$\int \cos x \, dx = \sin x + c$

$\int e^x \, dx = e^x + c$

$\int -\frac{1}{x^2} \, dx = \frac{1}{x} + c$

$\int 3 + 4x + x^2 \, dx = 3x + 2x^2 + \frac{1}{3}x^3 + c$

The constant c is included because the derivative of a constant is zero.

In general, $\int f(x) \, dx = F(x) + c$ means $F'(x) = f(x)$ for all x.

Worksheet 1

Find the following integrals:

1. a) $\int x \, dx$ b) $\int x^2 \, dx$ c) $\int x + x^2 \, dx$

    d) $\int (x-1)^2 \, dx$ e) $\int (x-2)^2 \, dx$ f) $\int (2x-1)^2 \, dx$

g) $\int (x-1)(x-2)\,dx$

h) $\int \left(\dfrac{2x+1}{x} \right) dx$

i) $\int \left(\dfrac{2x+1}{x^2} \right) dx$

j) $\int \sin x\,dx$

k) $\int e^{-x}\,dx$

l) $\int e^{2x}\,dx$

m) $\int \cos 2x\,dx$

n) $\int \sqrt{x}\,dx$

o) $\int \sqrt{1+x}\,dx$

2. If the slope of a function $y = f(x)$ is $x+2$ for all points on the graph of the function and if $f(3) = 11$, find $f(2)$.

3. A stone is dropped from rest down a well so that its acceleration due to gravity is (32 ft/sec) per second. Find how far the stone has fallen in 2 seconds.

4. A function $y = f(x)$ is such that at any point $\dfrac{dy}{dx} = \dfrac{1}{x+1}$. If the graph of the function passes through the point $(0,1)$ find $f(e-1)$.

Answers to Worksheet 1

1. a) $\dfrac{1}{2}x^2 + c$

b) $\dfrac{1}{3}x^3 + c$

c) $\dfrac{1}{2}x^2 + \dfrac{1}{3}x^3 + c$

d) $\dfrac{1}{3}(x-1)^3 + c$

e) $\dfrac{1}{3}(x-2)^3 + c$

f) $\dfrac{4}{3}x^3 - 2x^2 + x + c$

g) $\dfrac{1}{3}x^3 - \dfrac{3}{2}x^2 + 2x + c$

h) $2x + \ln x + c$

i) $2\ln x - \dfrac{1}{x} + c$

j) $-\cos x + c$

k) $-e^{-x} + c$

l) $\dfrac{1}{2}e^{2x} + c$

m) $\dfrac{1}{2}\sin 2x + c$

n) $\dfrac{2}{3}x^{\frac{3}{2}} + c$

o) $\dfrac{2}{3}(1+x)^{\frac{3}{2}} + c$

2. $f(2) = 6\dfrac{1}{2}$

3. 64 ft

4. 2

The process of integration is difficult because, although some learned techniques can be helpful, the only true test for verifying integration is to think backwards.

For example $\int \sec^2 x \, dx = \tan x + c$. This can only be found by knowing that $D(\tan x) = \sec^2 x$. Despite this, some basic techniques are helpful.

For example $\int x^{n-1} \, dx = \frac{1}{n} x^n + c$

because $\qquad D(x^n) = nx^{n-1}$.

e.g. $\qquad \int x^3 \, dx = \frac{1}{4} x^4 + c$.

Notice however that, even here, the "rule" breaks down for the isolated case where $n = 0$ since $\int x^{-1} = \ln|x| + c$.

$D(\ln x) = \frac{1}{x}$ is correct because, when x is negative, $\ln x$ is not defined and hence the derivative of $\ln x$ is similarly not defined.

However $\int \frac{1}{x} \, dx = \ln x + c$ clearly has 'problems' when x is negative. This can be avoided by establishing that $\int \frac{1}{x} \, dx = \ln|x| + c$.

If x is known to be positive throughout the range of interest in the question then it is usually best to eliminate the absolute value symbol.

Chain Rule Backwards

When trying to integrate $2x(x^2+1)^2$ i.e. $\int 2x(x^2+1)^2\,dx$, it is helpful to notice that $2x$

is the derivative of x^2+1. This means that $\int 2x(x^2+1)^2\,dx$ is an example of the Chain

Rule backwards because $D\left[\dfrac{1}{3}(x^2+1)^3\right]=\dfrac{1}{3}\cdot 3(x^2+1)^2\cdot 2x=2x(x^2+1)^2$.

i.e. $\qquad \int 2x(x^2+1)^2\,dx=\dfrac{1}{3}(x^2+1)^3+c$.

Similarly $\qquad \int \cos x(\sin x+1)^5\,dx$ can be considered in the same way since $\cos x$ is

the derivative of $(\sin x+1)$. Hence $\int \cos x(\sin x+1)^5\,dx=\dfrac{1}{6}(\sin x+1)^6+c$.

Another example would be $\int \dfrac{(\ln x)^2}{x}\,dx$ since $\int \dfrac{(\ln x)^2}{x}\,dx$ can be thought of as

$\int \dfrac{1}{x}(\ln x)^2\,dx$. Remember that $D(\ln x)=\dfrac{1}{x}$. Hence $\int \dfrac{1}{x}(\ln x)^2\,dx=\dfrac{1}{3}(\ln x)^3+c$.

Notice also that $\int x\sqrt{x^2+4}\,dx$ can be thought of as an example of the Chain Rule

backwards because "in essence", x is the derivative of x^2.

i.e. $\qquad \int x\sqrt{x^2+4}\,dx=\int x(x^2+4)^{\frac{1}{2}}\,dx=a(x^2+4)^{\frac{3}{2}}+c$ where a is some <u>constant</u>.

Let's find the value of a.

It is easy to differentiate $a(x^2+4)^{\frac{3}{2}}$ since $D\left[a(x^2+4)^{\frac{3}{2}}\right]=\dfrac{3}{2}a(x^2+4)^{\frac{1}{2}}\cdot 2x$

$$=3ax(x^2+4)^{\frac{1}{2}}$$

Since we are trying to find the value of a so that $D\left[a\left(x^2+4\right)^{\frac{3}{2}}\right] = x\left(x^2+4\right)^{\frac{1}{2}}$ it

follows that $3a=1$ i.e. $a=\dfrac{1}{3}$.

$$\therefore \qquad \int x\sqrt{x^2+4}\ dx = \frac{1}{3}\left(x^2+4\right)^{\frac{3}{2}}+c$$

It is highly recommended that, when you are trying to integrate something like

$7x^2\left(x^3+2\right)^{\frac{3}{5}}$ that you do <u>not</u> focus your attention on the constant of multiplication.

Instead concentrate on the fact that the "basic" solution to $\int 7x^2\left(x^3+2\right)^{\frac{3}{5}}dx$ is

$\left(x^3+2\right)^{\frac{8}{5}}$ since $D\left(x^3+2\right)^{\frac{8}{5}} = \dfrac{8}{5}\left(x^3+2\right)^{\frac{3}{5}}3x^2$ ✳

i.e. technique suggested is:

$$\text{Let } \int 7x^2\left(x^3+2\right)^{\frac{3}{5}}\ dx = a\left(x^3+2\right)^{\frac{8}{5}}$$

From ✳ we know that $7 = \dfrac{8}{5}a\cdot 3$

$$\therefore \qquad a = \frac{35}{24}$$

$$\therefore \qquad \int 7x^2\left(x^3+2\right)^{\frac{3}{5}}\ dx = \frac{35}{24}\left(x^3+2\right)^{\frac{8}{5}}+c.$$

Other helpful hints are that, for algebraic expressions only, if it is possible to divide a

denominator into a numerator then usually it is a good idea to do so.

e.g. $\quad \displaystyle\int\frac{x}{x-1}\ dx = \int 1+\frac{1}{x-1}\ dx = x+\ln|x-1|+c$

Also $\int \dfrac{x^3}{x^2+1}\,dx = \int x - \dfrac{x}{x^2+1}\,dx = \dfrac{1}{2}x^2 - \dfrac{1}{2}\ln\left|x^2+1\right| + c$

This hint does not necessarily apply to trigonometric fractions or expressions involving e^x.

e.g. $\int \dfrac{e^x}{e^x+1}\,dx = \ln\left|e^x+1\right| + c$ using the Chain Rule backwards.

Worksheet 2

1. Integrate the following:

   a) $\int 2x\left(x^2+1\right)^7 dx$

   b) $\int 2x\sin\left(x^2\right) dx$

   c) $\int 2x\sqrt{x^2-3}\, dx$

   d) $\int x\sqrt{9-4x^2}\, dx$

   e) $\int x^2\left(x^3+1\right)^2 dx$

   f) $\int x\left(x^3+1\right)^2 dx$

2. Integrate the following:

   a) $\int \sec^2 3x\, dx$

   b) $\int \dfrac{2x}{x^2+1}\, dx$

   c) $\int \dfrac{2x}{\left(x^2+1\right)^2}\, dx$

   d) $\int \dfrac{x^2}{x^3+1}\, dx$

   e) $\int \sin x \cos^7 x\, dx$

   f) $\int e^x\sqrt{e^x+1}\, dx$

   g) $\int \sqrt{e^{3x}+e^{2x}}\, dx$

   h) $\int \dfrac{e^x}{e^x+1}\, dx$

3. Integrate the following:

a) $\displaystyle\int \frac{x+1}{x-1}\, dx$

b) $\displaystyle\int \tan x\, dx$ (hint: write $\tan x$ as a fraction)

c) $\displaystyle\int xe^{-x^2}\, dx$

d) $\displaystyle\int \frac{1}{1+e^{-x}}\, dx$

e) $\displaystyle\int \frac{8}{\sqrt{3+4x}}\, dx$

f) $\displaystyle\int \frac{2x^2+4x+3}{x^2+x+1}\, dx$

g) $\displaystyle\int \frac{x}{x+1}\, dx$

h) $\displaystyle\int \frac{1}{x^2+6x+9}\, dx$

i) $\displaystyle\int \sqrt{x}\,(1+x)\, dx$

j) $\displaystyle\int 2^x\, dx$

k) $\displaystyle\int \frac{1}{\cos^2 x}\, dx$

Answers to Worksheet 2 (it is understood that each answer contains +c also)

1. a) $\dfrac{1}{8}\left(x^2+1\right)^8$

b) $-\cos\left(x^2\right)$

c) $\dfrac{2}{3}\left(x^2-3\right)^{\frac{3}{2}}$

d) $-\dfrac{1}{12}\left(9-4x^2\right)^{\frac{3}{2}}$

e) $\dfrac{1}{9}\left(x^3+1\right)^3$

f) $\dfrac{1}{8}x^8+\dfrac{2}{5}x^5+\dfrac{x^2}{2}$

2. a) $\dfrac{1}{3}\tan 3x$

b) $\ln\left|x^2+1\right|$

c) $-\left(x^2+1\right)^{-1}$

d) $\dfrac{1}{3}\ln\left|x^3+1\right|$

e) $-\dfrac{1}{8}\cos^8 x$

f) $\dfrac{2}{3}\left(e^x+1\right)^{\frac{3}{2}}$

g) same answer as for 2f) $\dfrac{2}{3}\left(e^x+1\right)^{\frac{3}{2}}$

h) $\ln\left|e^x+1\right|$

3. a) $x+2\ln|x-1|$

b) $-\ln|\cos x|$

c) $-\dfrac{1}{2}e^{-x^2}$

d) $\ln\left|e^x+1\right|$

e) $4\left(3+4x\right)^{\frac{1}{2}}$

f) $2x+\ln\left|x^2+x+1\right|$

g) $x-\ln|x+1|$

h) $-\left(x+3\right)^{-1}$

i) $\dfrac{2}{3}x^{\frac{3}{2}}+\dfrac{2}{5}x^{\frac{5}{2}}$

j) $\dfrac{1}{\ln 2}2^x$

k) $\tan x$

Worksheet 3

INTEGRATION

1. Find   a) $\int \dfrac{x}{3}\, dx$ b) $\int \cos x\,(1+\sin x)^4\, dx$

2. A stone is thrown horizontally at 12 feet per second from the top of a tower 100 feet high. When the stone is at point P (x,y) after t seconds $\dfrac{d^2 x}{dt^2}=0$

   and $\dfrac{d^2 y}{dt^2}=-32$.

   a) Express x in terms of t and y in terms of t.

   b) How long does the stone take to reach the ground?

   c) How far from the tower does it strike the ground?

3. Show that $v\dfrac{dv}{ds}$ is equivalent to acceleration where v is velocity and s is position. If a particle moves such that the acceleration $=-s$ and if the initial velocity is 10 metres per second find the position, from the starting point zero, where the particle first comes to rest.

4. Find: a) $\int x^2 \left(1+x^3\right)^6 dx$ b) $\int x^3 \left(2+x^4\right)^{-3} dx$ c) $\int \dfrac{1}{x^2}\left(1+\dfrac{1}{x}\right)^3 dx$

   d) $\int 2e^x \left(e^x+3\right)^4 dx$ e) $\int \dfrac{1}{2x}\, dx$ f) $\int \dfrac{2x^2}{3x^3+1}\, dx$

   g) $\int x\left(x^2+1\right)^3 dx$ h) $\int \dfrac{x+1}{x}\, dx$ i) $\int \dfrac{x}{x+1}\, dx$

   j) $\int \dfrac{x-1}{x+1}\, dx$ k) $\int \dfrac{1}{3x-2}\, dx$

l) $\int \cot x \, dx$ (hint: write $\cot x$ as a fraction)

m) $\int \cos 2x \, dx$

n) $\int e^{2x} \, dx$

o) $\int e^{\ln x^2} \, dx$

p) $\int (u^3 + 3)^3 u^2 \, du$

q) $\int (u^2 - 1)^4 u \, du$

r) $\int 7x (x^2 - 1)^{\frac{5}{6}} \, dx$

s) $\int \cos^2 x dx$

t) $\int \sin^2 x dx$

u) $\int xe^{-x^2} \, dx$

v) $\int \dfrac{e^{x^{-3}}}{x^4} \, dx$

w) $\int \dfrac{x+1}{x^2 + 2x + 2} \, dx$

5. Find the following integrals:

a) $4\int \dfrac{x^2}{x-1} \, dx$

b) $\int 3x \sin x^2 \, dx$

c) $\int \dfrac{1}{\sqrt{x}} (2\sqrt{x} + 1) dx$

d) $\int \dfrac{x}{x^2 + 1} \, dx$

e) $\int \dfrac{2x^2 - x}{x^3} \, dx$

f) $\int \dfrac{1}{\sqrt{2x-5}} \, dx$

6. If $\dfrac{dy}{dx} = \dfrac{2x}{x^2 + 1}$ and $y = 1$ when $x = 0$ find y when $x = 1$.

7. A particle is projected from a point O and t seconds later its velocity is $t - t^2$ feet per second. Find the time taken for the particle to return to O. Find the greatest distance of the particle from O during this time.

8. If the acceleration of the particle is given by $a = \sqrt{t+4}$ and it starts from rest find the velocity after 12 seconds.

Answers to Worksheet 3

1. a) $\dfrac{1}{6}x^2$ b) $\dfrac{1}{5}(1+\sin x)^5$

2. a) $x=12t;\quad y=100-16t^2$ b) $2\dfrac{1}{2}$ seconds   c) 30 feet

3. 10 feet

4. a) $\dfrac{1}{21}\left(1+x^3\right)^7+c$ b) $-\dfrac{1}{8}\left(2+x^4\right)^{-2}+c$ c) $-\dfrac{1}{4}\left(1+\dfrac{1}{x}\right)^4+c$

    d) $\dfrac{2}{5}\left(e^x+3\right)^5+c$ e) $\dfrac{1}{2}\ln|x|+c$ f) $\dfrac{2}{9}\ln\left|3x^3+1\right|+c$

    g) $\dfrac{1}{8}\left(x^2+1\right)^4+c$ h) $x+\ln|x|+c$ i) $x-\ln|x+1|+c$

    j) $x-2\ln|x+1|+c$ k) $\dfrac{1}{3}\ln|3x-2|+c$ l) $\ln|\sin x|+c$

    m) $\dfrac{1}{2}\sin 2x+c$ n) $\dfrac{1}{2}e^{2x}+c$ o) $\dfrac{1}{3}x^3+c$

    p) $\dfrac{1}{12}\left(u^3+3\right)^4+c$ q) $\dfrac{1}{10}\left(u^2-1\right)^5+c$ r) $\dfrac{21}{11}\left(x^2-1\right)^{\frac{11}{6}}+c$

    s) $\dfrac{1}{2}x+\dfrac{1}{4}\sin 2x+c$ t) $\dfrac{1}{2}x-\dfrac{1}{4}\sin 2x+c$ u) $-\dfrac{1}{2}e^{-x^2}+c$

    v) $-\dfrac{1}{3}e^{\frac{1}{x^3}}+c$ w) $\dfrac{1}{2}\ln\left(x^2+2x+2\right)+c$

5. a) $2x^2+4x+4\ln(x-1)+c$ b) $-\dfrac{3}{2}\cos x^2+c$ c) $\dfrac{1}{2}\left(2\sqrt{x}+1\right)^2+c$

    d) $\dfrac{1}{2}\ln\left(x^2+1\right)+c$ e) $2\ln|x|+\dfrac{1}{x}+c$ f) $(2x-5)^{\frac{1}{2}}+c$

6. $1+\ln 2$ 7. $1\dfrac{1}{2}$ secs. $\dfrac{1}{6}$ foot.   8. $37\dfrac{1}{3}$

CHAPTER 13

Definite Integrals

Since integration can be used in a practical sense in many applications it is often useful to have integrals evaluated for different values of the variable of integration. Frequently we wish to integrate an expression between some limits. The practical significance of this will be explained later in this chapter.

For example, if we wish to integrate $x^2 + 2x$ between $x = 1$ and $x = 3$ then we write

$\int_1^3 x^2 + 2x \ dx$ which is $\left[\dfrac{x^3}{3} + x^2 \right]_1^3$.

Mathematically this means find the value of $\dfrac{x^3}{3} + x^2$ when $x = 3$ and then <u>subtract</u> the value of $\dfrac{x^3}{3} + x^2$ when $x = 1$.

i.e.
$$\int_1^3 x^2 + 2x \ dx = \left[\dfrac{x^3}{3} + x^2 \right]_1^3 = (9+9) - \left(\dfrac{1}{3} + 1 \right) = 16\dfrac{2}{3}.$$

Similarly
$$\int_{-2}^4 3x^2 \ dx = \left[x^3 \right]_{-2}^4 = 4^3 - \left(-2^3 \right) = 72$$

And also
$$\int_1^2 \dfrac{x+1}{x} \ dx = \int_1^2 \left(1 + \dfrac{1}{x} \right) dx = \left[x + \ln x \right]_1^2$$

$$= (2 + \ln 2) - (1 + \ln 1) = \underline{1 + \ln 2}.$$

In general $\int_a^b f(x) \ dx = F(b) - F(a)$ where $F'(x) = f(x)$.

This is called the **Fundamental Theorem of Calculus**.

Note that it is not necessary to include the constant of integration since the subtraction "cancels out" that constant.

Such integrals are called DEFINITE INTEGRALS because we are substituting definite values of x.

Worksheet 1

DEFINITE INTEGRALS

1. Evaluate the following definite integrals. You may use a calculator.

a) $\int_{1}^{\sqrt{5}} \dfrac{2x}{\sqrt{x^2-1}}\, dx$

b) $\int_{0}^{\frac{\pi}{8}} \sin 4x\, dx$

c) $\int_{0}^{1} \dfrac{2}{x+1}\, dx$

d) $\int_{0}^{1} \dfrac{2x}{1+x^2}\, dx$

e) $\int_{0}^{1} \dfrac{2}{\sqrt{1+x}}\, dx$

f) $\int_{0}^{1} \dfrac{1}{x^2+6x+9}\, dx$

g) $\int_{0}^{1} \dfrac{x}{x+1}\, dx$

h) $\int_{0}^{1} \dfrac{8}{3+4x}\, dx$

i) $\int_{0}^{1} \dfrac{8x}{3+4x}\, dx$

j) $\int_{0}^{1} \dfrac{8}{\sqrt{3+4x}}\, dx$

k) $\int_{0}^{1} \dfrac{2x^2+4x+3}{x^2+x+1}\, dx$

l) $\int_{0}^{1} \dfrac{e^x}{1+e^x}\, dx$

m) $\int_{0}^{\frac{\pi}{4}} e^{\cos x} \cdot \sin x\, dx$

n) $\int_{0}^{1} x^2 \sqrt{1-x^3}\, dx$

o) $\int_{0}^{e-1} \dfrac{2x+1}{x+1}\, dx$

p) $\int_{0}^{\pi} \sin 3x\, dx$

q) $\int_{0}^{3} \dfrac{2}{(x+1)^2}\, dx$

r) $\int_{\frac{\pi}{4}}^{\frac{\pi}{2}} 1+\sin x\, dx$

s) $\int_{\frac{\pi}{4}}^{\frac{\pi}{2}} \sin^2 x \cos x \, dx$

t) $\int_{0}^{\frac{\pi}{2}} \cos x e^{\sin x} \, dx$

u) $\int_{0}^{\frac{\pi}{4}} \frac{\sec^2 x}{1 + \tan^2 x} \, dx$

v) $\int_{0}^{2} \frac{x^3 + 2x}{x^2 + 1} \, dx$

w) $\int_{0}^{\frac{1}{2}} 2x\sqrt{1 - 4x^2} \, dx$

Answers to Worksheet 1

1. a) 4

b) 0.25

c) 1.3863

d) 0.693

e) 1.657

f) $\frac{1}{12}$

g) 0.30685

h) 1.6946

i) 0.729

j) 3.655

k) 3.0986

l) 0.62

m) 0.6902

n) $\frac{2}{9}$

o) 2.43656

p) $\frac{2}{3}$

q) $1\frac{1}{2}$

r) 1.4925

s) 0.2155

t) 1.718

u) $\frac{\pi}{4}$

v) 2.80472

w) $\frac{1}{6}$

Worksheet 2

1. Evaluate the following integrals without using a calculator.

   a) $4\int_{2}^{3} \dfrac{x^2}{x-1}\, dx$ b) $\int_{0}^{\sqrt{\pi}} 3x\sin\left(x^2\right) dx$

   c) $\int_{0}^{\sqrt{e-1}} \dfrac{x}{x^2+1}\, dx$ d) $\int_{1}^{e} \dfrac{2x^2 - x}{x^3}\, dx$

   e) $\int_{1}^{4} \dfrac{1}{x^2\sqrt{x}}\, dx$ f) $\int_{1}^{2} e^{\ln x+2}\, dx$

2. $\dfrac{dy}{dx} = 3x^2$ and $y = 5$ when $x = 2$. Find y when $x = 3$.

3. $\dfrac{dy}{dx} = x+1$ and $y = \dfrac{1}{2}$ when $x = 3$. Find y when $x = 4$.

4. A particle starts with an initial velocity of 3 feet per second. Its acceleration is $(3t+1)$ feet/sec², where t is the number of seconds from the start. Find the velocity after 2 seconds.

5. Evaluate the following integrals without a calculator:

   a) $\int_{1}^{\log_2 3} 2^x\, dx$ b) $\int_{0}^{e-1} \dfrac{x+2}{x+1}\, dx$

   c) $\int_{0}^{1} \dfrac{x^2+2x}{(x+1)^2}\, dx$

Answers to Worksheet 2

1. a) $14 + 4\ln 2$ b) 3 c) $\dfrac{1}{2}$ d) $1 + \dfrac{1}{e}$ e) $\dfrac{7}{12}$ f) $\dfrac{3e^2}{2}$

2. 24   3. 5   4. 11   5. a) $\dfrac{1}{\ln 2}$ b) e c) $\dfrac{1}{2}$

Applications of Definite Integrals to Area

Example 1

Question: Find the area between the x-axis, the graph of $y = 2x$ and $x = 4$.

Answer:

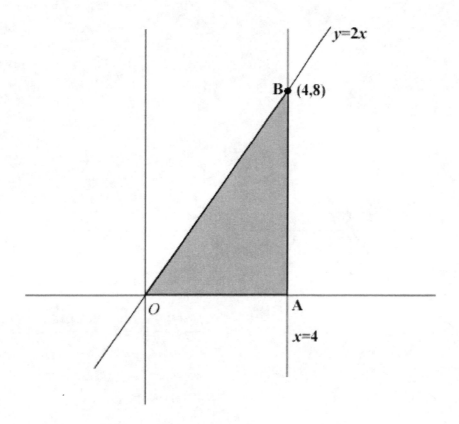

We are trying to find the area of ΔOAB.

By elementary geometry: Area $\Delta OAB = \dfrac{OA \times AB}{2} = \dfrac{4 \times 8}{2} = 16$.

Note also that $\displaystyle\int_0^4 2x \; dx = \left[x^2 \right]_0^4 = 4^2 - 0^2 = 16$.

Example 2

Question: Find the area bounded by $y = 2x$, $x = 3$, $x = 7$, and the x-axis.

Answer:

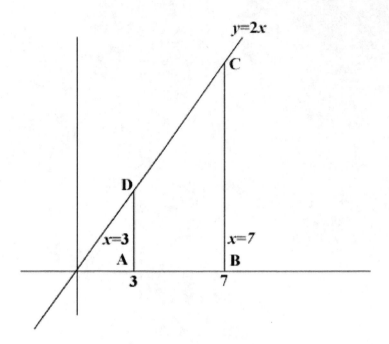

We wish to find the area of $ABCD = \left(\dfrac{AD + BC}{2}\right) \times AB = \left(\dfrac{6 + 14}{2}\right) \times 4 = 40$.

Now consider $\displaystyle\int_{3}^{7} 2x \; dx = \left[x^2\right]_{3}^{7} = 49 - 9 = 40$ also.

It appears as though area is related to the definite integral.

Area Under a Curve as a Definite Integral

Let $f(x)$ be a positive continuous function as shown below.

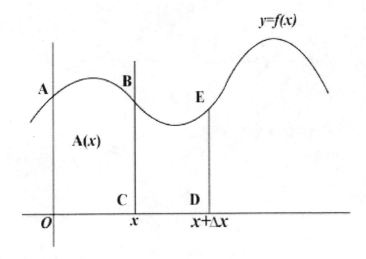

We will try to find area under the curve bounded by the y-axis, $y = f(x)$ and the

x-axis from 0 to x i.e. we are trying to find Area $O\overset{\frown}{A}BC$.

Let $A(x)$ represent the area. To evaluate $A(x)$ for different values of x it is helpful

to investigate the derivative of $A(x)$.

By First Principles, the derivative of $A(x)$

$$= A'(x) = \lim_{\Delta x \to 0} \frac{A(x + \Delta x) - A(x)}{\Delta x}$$

$$= \lim_{\Delta x \to 0} \frac{\text{Area } O\overset{\frown}{A}ED - \text{Area } O\overset{\frown}{A}BC}{\Delta x}$$

$$= \lim_{\Delta x \to 0} \frac{\text{Area } C\overset{\frown}{B}ED}{\Delta x} \quad \circledast$$

Now consider the region $C\overset{\frown}{B}ED$.

Remember that $CD = \Delta x$ is considered as a "small" change in x.

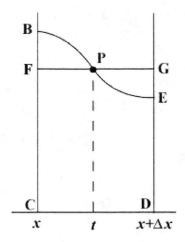

If we think of $C\overset{\frown}{B}ED$ as a body of sand whose upper edge is $\overset{\frown}{BE}$ then it is clear that

$\overset{\frown}{BE}$ can be "smoothed out" horizontally so that area of $C\overset{\frown}{B}ED$ = area of rectangle

$CFGD$ (see above). This is true also by the Intermediate Value Theorem. It is also

clear that FG intersects $\overset{\frown}{BE}$ at some point P whose x co-ordinate lies between x

and $x + \Delta x$.

Let P be $(t, f(t))$ where $x \leq t \leq x + \Delta x$.

$$\text{Area } C\overset{a}{\overset{\frown}{B}}ED = \text{ Area } CFGD = CD \times (y \text{ co-ordinate of } P) = \Delta x \cdot f(t).$$

Substituting into ✳ on the previous page, it follows that

$$A'(x) = \lim_{x \to 0} \frac{\Delta x \cdot f(t)}{\Delta x} = \lim_{\Delta x \to 0} f(t)$$

In the limiting case as $\Delta x \to 0$, $t \to x$.

$\therefore \quad \underline{\underline{A'(x) = f(x)}}$. It follows that $A(x) = \int f(x)\, dx$.

To find the area under a curve $y = f(x)$, bounded by $x = a$, $x = b$ and the x-axis we note that this is represented by region S in the diagram:

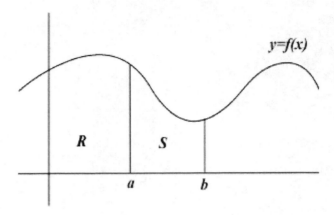

Area of region S is therefore $\int_a^b f(x)\ dx$.

Comment

Students often think that $\int_0^a f(x)\ dx$ is the area of region R and $\int_0^b f(x)\ dx$ is the area of region $R + S$ but strictly speaking this is not true.

$$\text{Area region } R = \int_0^a f(x)\ dx + k \text{ for some constant } k$$

$$\text{And similarly Area region } R + S = \int_0^b f(x)\ dx + \text{ the same constant } k$$

It does however remain true that

$$\text{Area } S = \int_a^b f(x)\ dx \text{ since the constant } k \text{ "cancels out".}$$

Example

Find the area under $y = x^2$ from $x = 2$ to $x = 4$ above the x-axis.

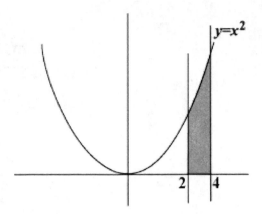

$$\text{Area} = \int_2^4 x^2 \, dx = \left[\frac{x^3}{3}\right]_2^4 = \frac{64}{3} - \frac{8}{3} = \frac{56}{3} = 18\frac{2}{3}$$

Example

Find the area bounded by $y = \dfrac{1}{1+x^2}$, $x = -1$, $x = \sqrt{3}$ and the x-axis i.e. find

the area of shaded region R in the picture below:

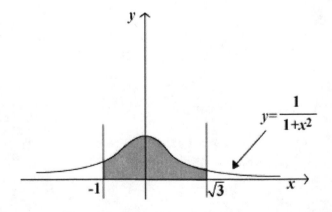

$$\text{Area} = \int_{-1}^{\sqrt{3}} \frac{1}{1+x^2}\, dx = \left[\, \text{Arctan} x \,\right]_{-1}^{\sqrt{3}}$$

$$= \text{Arctan}\sqrt{3} - \text{Arctan}(-1)$$

$$= \frac{\pi}{3} - \left(-\frac{\pi}{4}\right)$$

$$= \frac{7\pi}{12}.$$

To find the area of a region whose boundaries involve more complex curves it is

helpful to consider a thin strip procedure as follows.

Example

Find the area bounded by $y = 2x$ and $y = 2x^2$.

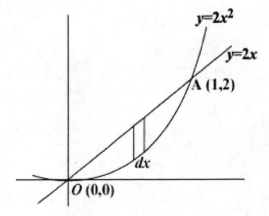

Note that the intersection points O and A are (0,0) and (1,2) respectively.

Draw a thin vertical strip in the region whose area is to be evaluated.

Consider the thin strip as though it were a rectangle whose area is

$$\left[\left(y \text{ co-ordinate on } y = 2x \right) - \left(y \text{ co-ordinate on } y = 2x^2 \right) \right] \text{times } dx$$

$$= \left(Y - y \right) dx$$

$$= \left(2x - 2x^2 \right) dx$$

To evaluate the area of the region we need to add up an infinite number of thin strips whose width dx tends to the limit of 0. This is effected by

$$\int_0^1 2x - 2x^2 \ dx \qquad \text{because integration is a process of adding an}$$

infinite number of values.

Note that the thin strips vary from $x = 0$ to $x = 1$ which are the limits of integration for the variable x. Note also that dx has changed its role from width of the strip to the variable of integration for reasons beyond the scope of this text.

$$\text{Area of required region} \quad = \int_0^1 \left(2x - 2x^2 \right) dx$$

$$= \left[x^2 - \frac{2}{3} x^3 \right]_0^1 = \left(1 - \frac{2}{3} \right) - \left(0 - 0 \right)$$

$$= \frac{1}{3}$$

Note that in a later chapter a similar process will be used to evaluate volumes.

Sometimes it is more convenient (and even perhaps required) to draw thin horizontal strips to evaluate an area.

Example

Find the area (in the first quadrant) bounded by $y = \dfrac{1}{x}$, $y = x^2$, and $y = 4$.

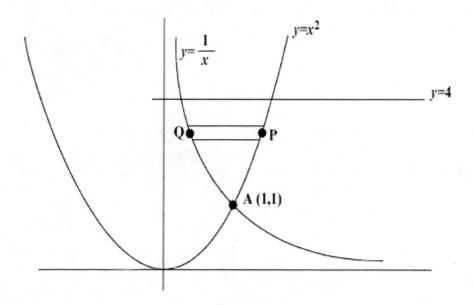

Note that point A, the intersection of $y = x^2$ and $y = \dfrac{1}{x}$ is $(1,1)$.

In this example it is better to draw a thin horizontal strip because using vertical strips would necessitate that the region be divided into two separate parts.

The area of the thin horizontal strip is $(X - x)dy$ where X denotes the x co-ordinate of point P on $y = x^2$ and x denotes the x co-ordinate of point Q on $y = \dfrac{1}{x}$.

Area of region is
$$\int_1^4 (X - x)\, dy = \int_1^4 \sqrt{y} - \frac{1}{y}\, dy$$

$$= \left[\frac{2}{3} y^{\frac{3}{2}} - \ln y \right]_1^4 = \left(\frac{16}{3} - \ln 4 \right) - \left(\frac{2}{3} - \ln 1 \right)$$

$$= \frac{14}{3} - \ln 4 = 3.28 \text{ (approx.)}$$

Note also that the thin strips vary from $y = 1$ to $y = 4$ and hence are the limits of integration.

It is sometimes assumed that area under a curve is simply obtained by integrating the function. Care however and understanding are required as illustrated by the following example.

Example

Consider the function $f(x) = 4x^3 - 24x^2 + 64 = 4(x - 2)(x^2 - 4x - 8)$

as shown below.

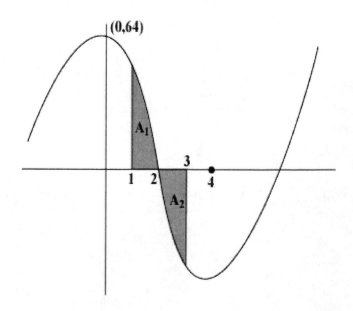

Let's obtain the area $A_1 + A_2$ i.e. the area on either side of the x-axis from $x = 1$ to $x = 3$.

If we evaluate $\int_1^3 4x^3 - 24x^2 + 64 \; dx$ we get

$$\left[x^4 - 8x^3 + 64x\right]_1^3 = (81 - 216 + 192) - (1 - 8 + 64)$$

$$= (57) - (57)$$

$$= 0$$

It is clear that the area of region $A_1 + A_2$ is not zero.

In fact for the function chosen, (2,0) is the point of inflection and the function is symmetric about (2,0). In this case area A_1 = area A_2.

Another example follows which illustrates how to calculate areas of regions similar to those in the preceding example.

Example

Find the area, both above and below the x-axis bounded by the x-axis and a graph of $y = x(x - 2)(x - 4)$.

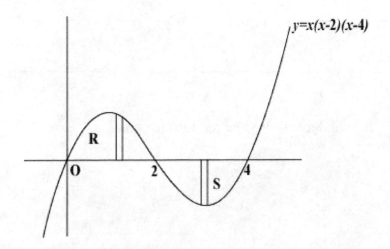

Note that it is not correct to say that the area $R+S = \int_0^4 y \, dx$ because in region

R the thin strip has height y whereas in region S the thin strip has a height

of $-y$. It must be remembered that x and y refer to the x and y co-

ordinates of a point on the graph.

The area of region R + the area of region S

$$= \int_0^2 y \, dx + \int_2^4 \left(-y \right) dx$$

$$= \int_0^2 x^3 - 6x^2 + 8x \, dx + \int_2^4 -x^3 + 6x^2 - 8x \, dx$$

$$= \left[\frac{x^4}{4} - 2x^3 + 4x^2 \right]_0^2 + \left[-\frac{x^4}{4} + 2x^3 - 4x^2 \right]_2^4$$

$$= \left[\left(4 - 16 + 16 \right) - \left(0 - 0 + 0 \right) \right] + \left[\left(-64 + 128 - 64 \right) - \left(-4 + 16 - 16 \right) \right]$$

$$= 4 + 4 = 8$$

Note that, using a graphing calculator, the area can be evaluated by

$$\int_0^4 \left| x^3 - 6x^2 + 8x \right| dx \, .$$

Sometimes a region will have to be divided into two distinct parts.

Example

Find the area bounded by $y = 2^x$, $y = 4^x$ and $y = \dfrac{1}{x}$.

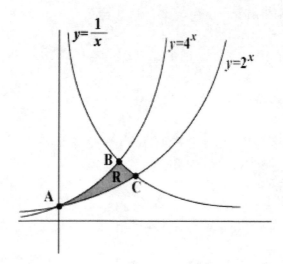

It is clear from the graph that the region R cannot be evaluated in <u>one</u> step by

using horizontal or vertical strips. Point A is $(0,1)$, point B $(\dfrac{1}{2},2)$ and point

C can be found by graphing calculator to be $(0.641, 1.56)$ (approx).

Considering thin vertical strips it follows that the area of region R

$$R = \int_{0}^{\frac{1}{2}} 4^x - 2^x \, dx + \int_{\frac{1}{2}}^{0.641} \frac{1}{x} - 2^x \, dx$$

$$= \left[\frac{4^x}{\ln 4} - \frac{2^x}{\ln 2} \right]_{0}^{\frac{1}{2}} + \left[\ln x - \frac{2^x}{\ln 2} \right]_{\frac{1}{2}}^{0.641}$$

$$= \left[\left(\frac{2}{\ln 4} - \frac{\sqrt{2}}{\ln 2} \right) - \left(\frac{1}{\ln 4} - \frac{1}{\ln 2} \right) \right] + \left[\left(\ln 0.641 - \frac{2^{0.641}}{\ln 2} \right) - \left(\ln \frac{1}{2} - \frac{\sqrt{2}}{\ln 2} \right) \right]$$

$$= \left(0.12376 \right) + \left(0.03895 \right) \text{ (using a calculator)}$$

$$= 0.16271 \text{ (approx.)}$$

Worksheet 3

1. Find the area bounded by the x-axis, the y-axis and $x+y=4$.

2. Find the area bounded by $y=x^2+1$, $x=2$, the x-axis and the y-axis.

3. Find the area under one half-period of $y=\sin x$.

4. Find the area under one half-period of $y=2\sin 3x$.

5. Find the area cut off by the x-axis, above the x-axis, from the parabola

   $y=(3+x)(4-x)$.

6. Find the area bounded by $y=4-x^2$, the x-axis, the y-axis, and $x=1$.

7. Find the area cut off both above and below the x-axis by

   $y=(x-1)(x-2)(x-3)$.

8. Find the area bounded by $y=\sin x$ and $y=\dfrac{1}{2}$ for $0\le x\le \pi$.

9. Find the area bounded by $y=\sin x$, $y=\dfrac{1}{2}$, $x=\dfrac{\pi}{3}$, and $x=\dfrac{2\pi}{3}$.

10. Find the area between $y=x^2$ and $y=8x$.

11. Find the area between $y=x^2$ and $y=x$.

12. Find the area bounded by $y=e^x$, $y=e^{-x}$, and $x=1$.

13. Find the area between the graphs of $y=\sin x$ and $y=\cos x$ between two

    consecutive points of intersection.

14. Find the area enclosed by the positive x-axis and the curve $y=3x^2-x^3$.

15. Find the smaller area bounded by $y=4x-x^3$ and $y=x^2-2x$.

16. Consider the region R in the first quadrant bounded by the x-axis, the y-axis, and $y = 4 - x^2$.

   i) Find the area of region R.

   ii) Find the value of k so that $x = k$ divides the region R into two equal regions of equal area.

   iii) Find the value of k so that $y = k$ divides the region R into two equal regions of equal area.

   iv) Find the value of k so that $y = kx$ divides the region R into two regions of equal area

17. Find the area between $y = x$, $y = \dfrac{1}{x^2}$, the x-axis and $x = 3$.

18. Find the area bounded by $y = x + 4$ and $y = x^2 + 2$.

19. Find the area between $y^2 = -4(x - 1)$ and $y^2 = -2(x - 2)$.

20. Find the area between the graphs of $y = 4 - x^2$, and $y = 2(x^2 - 4)$.

21. Find the area in the first quadrant bounded by the graphs of $y = x^{-2}$ and $y = 4\dfrac{1}{4} - x^2$.

22. Find the area lying above the x-axis of $y = \sin x + \cos x$ for $-\pi \le x \le \pi$.

23. Find the area bounded by $y = \dfrac{x+1}{x}$, $y = x^2 + 1$, and $x = 2$.

Answers to Worksheet 3

1. 8

2. $\dfrac{14}{3}$

3. 2

4. $\dfrac{4}{3}$

5. $57\dfrac{1}{6}$

6. $\dfrac{11}{3}$

7. $\dfrac{1}{2}$

8. $\sqrt{3} - \dfrac{\pi}{3}$

9. $1 - \dfrac{\pi}{6}$

10. $85\dfrac{1}{3}$

11. $\dfrac{1}{6}$

12. 1.086

13. $2\sqrt{2}$

14. $6\dfrac{3}{4}$

15. $5\dfrac{1}{3}$

16. i) $\dfrac{16}{3}$ ii) $k = 0.6946$

    iii) $k = 1.48$ iv) $k = 2.163$

17. $\dfrac{7}{6}$ 21. $2\dfrac{1}{4}$

18. $4\dfrac{1}{2}$ 22. $2\sqrt{2}$

19. $2\dfrac{2}{3}$ 23. 1.64

20. 32

Average Value

The average value of a finite set of data is of course the sum of the data divided by the number of items.

In the case of average value of a function over a range of values of x we are considering an infinite number of values and hence we use a graphical approach.

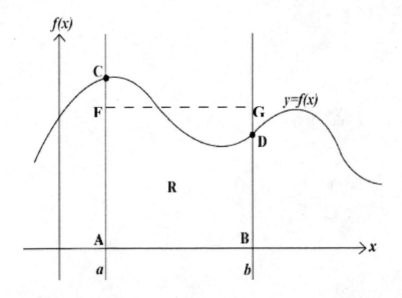

To find the average value of $f(x)$ as x varies between a and b we think of region R as though it were sand. The "wavy" top of the sand from C to D could be "smoothed out" horizontally so that the area of region R is re-arranged into a rectangle indicated by the dotted line.

i.e.　　area $AFGB$ = area $A\overset{a}{C}DB$

The height of the dotted line i.e. AF is the average value of the function as x varies from a to b.

$$\text{area } AFGB = \text{area } A\overset{a}{C}DB$$

$$\therefore \quad AF \times (b-a) = \int_{a}^{b} f(x) \, dx$$

$$\therefore \quad AF = (\text{average value}) = \frac{\int_{a}^{b} f(x) \, dx}{b-a}$$

Note that this result would apply even if $f(x)$ took on negative values.

Example

The average value of $\sin x$ as x varies between $x = \dfrac{\pi}{2}$ and $x = \pi$

$$= \frac{\int_{\frac{\pi}{2}}^{\pi} \sin x \, dx}{\pi - \frac{\pi}{2}}$$

$$= \frac{\left[-\cos x \right]_{\frac{\pi}{2}}^{\pi}}{\frac{\pi}{2}}$$

$$= \frac{1}{\frac{\pi}{2}}$$

$$= \frac{2}{\pi}$$

Worksheet 4

Areas Under A Curve/Average Value

1. Find the area bounded by $x = y(4-y)$ and $x + y = 4$.

2. Find the area between $y = x$, $y = \dfrac{1}{x^2}$, the x-axis and $x = 3$.

3. Find the average value of x^n as x varies from 0 to 1, where $n > 0$.

4. Find the area bounded by the parabola $x = 8 + 2y - y^2$, the y-axis, $y = -1$ and $y = 3$.

5. Find the area bounded by the parabola $y^2 = 4x$ and the line $y = 2x - 4$.

6. Find the area enclosed by the graphs $y = 6x - x^2$ and $y = x^2 - 2x$.

7. Find the average height of a semi-circle of radius 4.

8. Find the finite area enclosed by the three graphs $y = 4^x$, $y = 2^{x+1}$, and $y = 2^x + 6$.

9. Find the average value of x^3 as x varies between 2 and 4.

10. Find the average value of x^3 as x varies between -1 and 2.

11. Find the area bounded by the parabola $x = 1 + y^2$ and the line $x = 10$.

12. Find the average value of $\dfrac{1}{x^2}$ as x varies between $\dfrac{1}{2}$ and 1.

13. Find the average value of $\sin x$ as x varies between 0 and π.

14. Find the area in the first quadrant bounded by $y = 2^x$, $y = 3^x$ and $x = 2$.

15. Find the area bounded by the parabolas $x = 6y - y^2$ and $x = y^2 - 2y$.

16. Find the area bounded by $y = \sqrt{x}$, $y = \dfrac{1}{x^2}$, the x-axis and $x + y = \dfrac{9}{4}$.

17. Find the area bounded by $y = x^2 + 2x - 4$ and $y = -2x^2 + 4x - 3$.

18. Find the area bounded by $y^2 = x$ and $3y = 5x^2 - 8$ (use your graphing calculator).

19. Sketch $2y = x(x-4)(2x-5)$. The line $y = x$ cuts this curve at R and at A and B where A is between R and B. Find the area bounded by the curve from R to A and the line $y = x$.

Answers to Worksheet 4

1. $4\dfrac{1}{2}$

2. $1\dfrac{1}{6}$

3. $\dfrac{1}{n+1}$

4. $\dfrac{92}{3}$

5. 9

6. $\dfrac{64}{3}$

7. π

8. $6 - \dfrac{5}{\ln 4}$

9. 30

10. $\dfrac{5}{4}$

11. 36

12. 2

13. $\dfrac{2}{\pi} \approx 0.6366$

14. 2.9538

15. $\dfrac{64}{3}$

16. $\dfrac{17}{32}$

17. $\dfrac{32}{27}$

18. 1.907

19. $\dfrac{14}{3}$

Functions Defined As Definite Integrals

Sometimes functions are defined in terms of another integral. For example, $f(x)$ could equal

$$\int_2^x \left(1 + 3t^2\right)\, dt$$

This means that $\quad f(x) = \left[t + t^3\right]_2^x = x + x^3 - 10$

When the integral can be evaluated easily then $f(x)$ is defined explicitly. However $f(x)$ might be defined as (say)

$$\int_1^x \sqrt{t + t^2}\, dt$$

In this example $f(x)$ cannot easily be defined explicitly and hence graphing calculators, numerical methods, etc. need to be used to evaluate f for different values of x. However, note that it is possible to define $f'(x)$ explicitly.

In fact, $f'(x) = \sqrt{x + x^2}$.

This can be shown as follows:

Let $\int \sqrt{t + t^2}\, dt = F(t)$ for some function F where $F'(t) = \sqrt{t + t^2}$.

It follows that $\int_1^x \sqrt{t + t^2}\, dt = F(x) - F(1)$.

If we now differentiate both sides of this equation with respect to x we get

$$D_x \int_1^x \sqrt{t + t^2}\, dt = F'(x) - 0$$

$$= \sqrt{x + x^2}\ .$$

Note that the <u>numerical</u> value of the lower limit does not change $f'(x)$.

Similarly if $\quad g(x) = \int_2^{x^2} \sqrt{t + t^2}\ dt$

Then $\quad g'(x) = F'\left(x^2\right)$ times $2x$ (by Chain Rule)

$$= \sqrt{x^2 + x^4} \cdot 2x$$

Sometimes, $f(x)$ may be defined as (say)

$$f(x) = \int_1^x g(t)\ dt \text{ where } g(t) \text{ is as shown below.}$$

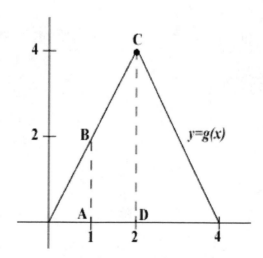

For example, $f(1) = \int_1^1 g(t)\, dt = 0$, $f(2) = \int_1^2 g(t)\, dt =$ Area $ABCD = 3$

and $\quad f(0) = \int_1^0 g(t)\, dt = -\int_0^1 g(t)\, dt = -$Area $\triangle ABO = -1$

Worksheet 5

1. Evaluate the following integrals without the aid of a calculator.

   a) $\displaystyle\int_0^{\ln 2} \frac{e^x}{e^x+1}\, dx$

   b) $\displaystyle\int_2^5 \frac{4}{2x-1}\, dx$

   c) $\displaystyle\int_0^2 x^2\left(1+x^3\right) dx$

   d) $\displaystyle\int_0^1 \frac{1+x^2}{1+x}\, dx$

2. Find the average value of $\sqrt{x^2-x^4}$ as x varies between 0 and 1.

3. Is the average value of $\sqrt{f(x)}$ as x varies between a and b the square root of the average value of $f(x)$ as x varies between a and b? Explain.

Calculators are permitted for the rest of the questions on Worksheet 5.

4. The graph of a differentiable function f on the closed interval [-6,6] is shown on the next page. Let $G(x) = \displaystyle\int_{-6}^x f(t)\, dt$ for $-6 \le x \le 6$.

   a) Find $G(-4)$.

   b) Find $G'(-1)$.

   c) On which interval or intervals is the graph of G concave down? Justify your answer.

   d) Find the value of x at which G has its maximum on the closed interval [-4,4].

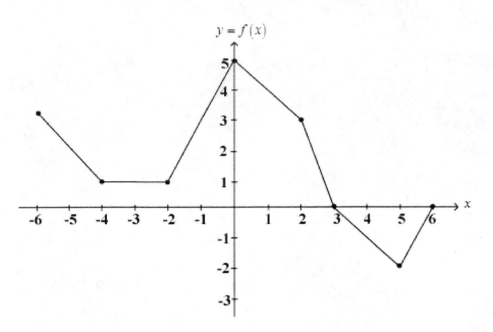

5. a) Find the area, A, as a function of k, of the region in the first quadrant

   enclosed by the y-axis and the graphs of $y = x^{\frac{1}{3}}$ and $y = k$ for $k > 0$.

   b) What is the value of A when $k = 2$?

   c) If the line $y = k$ is moving upward at the rate of $\dfrac{1}{2}$ units per second, at

   what rate is A changing when $k = 2$?

6. Suppose $\displaystyle\int_{2}^{4} f(x)\ dx = 6$ and $\displaystyle\int_{2}^{7} f(x)\ dx = 7$. Find $\displaystyle\int_{4}^{7} \left[f(x) + 4 \right] dx$.

7. Suppose that $\displaystyle\int_{1}^{7} f(x)\ dx = 19$, $\displaystyle\int_{1}^{5} f(x)\ dx = 11$ and $\displaystyle\int_{1}^{5} g(x)\ dx = 14$. Evaluate the

   following:

   a) $\displaystyle\int_{1}^{5} f(x) + g(x)\ dx$ \qquad\qquad b) $\displaystyle\int_{1}^{7} 5f(x)\ dx$

   c) $\displaystyle\int_{1}^{5} g(x) + 2\ dx$ \qquad\qquad d) $\displaystyle\int_{5}^{7} f(x)\ dx$

8. Let $F(x) = \int_{-2}^{x} f(t)\, dt$, $-4 \le x \le 4$, where f is the function whose graph is

shown below. The curved paths are semi-circles.

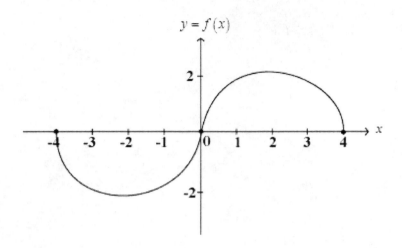

a) Complete the following table:

x	-4	-2	0	2	4
$F(x)$					

b) For what intervals is F concave down?

c) Does F have an inflection point? If so, at what value(s) of x?

d) In what interval(s) is F increasing?

9. Let $f(x) = \begin{cases} \dfrac{\sin x}{x} & \text{for } x \ne 0 \\ 1 & \text{for } x = 0 \end{cases}$ and $A(x) = \int_{0}^{x} f(t)\, dt$.

a) Sketch the graph of the function f.

b) At what values of x does the function A have a local minimum?

c) Find the coordinates of the first inflection point where $x > 0$ in the graph

of A.

10. The graph of a function f is shown.

Let $F(x) = \int_{-1}^{x} f(t)\, dt$.

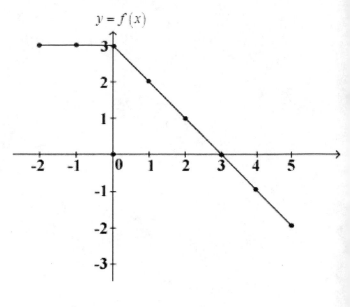

$y = f(x)$

a) Find $F(-1)$, $F(0)$, $F(3)$

   and $F(5)$.

b) Find $F(-2)$.

c) If $K(x) = \int_{-2}^{x} f(t)\, dt$,

   determine $K(x) - F(x)$.

d) When is the graph of F concave down?

Answers to Worksheet 5

1. a) $\ln\left(\dfrac{3}{2}\right)$ b) $\ln 9$ c) $\dfrac{40}{3}$ d) $\ln 4 - \dfrac{1}{2}$ 2. $\dfrac{1}{3}$

3. No since $\int_{a}^{b} \sqrt{f(x)}\, dx$ is not equal to $\sqrt{\int_{a}^{b} f(x)\, dx}$.

4. a) 4   b) 3   c) $-6 < x < -4$ or $0 < x < 5$ d) $x = 3$

5. a) $A = \dfrac{k^4}{4}$ b) 4   c) 4 square units/sec   6. 13

7. a) 25   b) 95   c) 22   d) 8

8. a)

x	-4	-2	0	2	4
$F(x)$	π	0	$-\pi$	0	π

b) $-4 < x < -2$ OR $2 < x < 4$ c) $x = -2$ or $x = 2$ d) $0 < x < 4$

9. a)

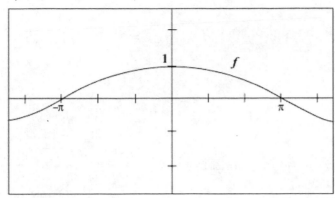

b) Since $A'(x) = f(x)$, A will have a local minimum when $A'(x) = f(x)$

changes from negative to positive. So there is a minimum at $x = -\pi$. But

also, when $x > 0$, $\dfrac{\sin x}{x}$ will have the desired sign changes at

$x = -\pi,\ -3\pi,\ -5\pi,\ ...,\ -(2k+1)\pi$, where k is any positive integer or

$x = 2\pi,\ 4\pi,\ ...,\ 2k\pi$, where k is any positive integer.

c) $A''(x) = \dfrac{d}{dx}\left(\dfrac{\sin x}{x}\right) = \dfrac{x\cos x - \sin x}{x^2}$. The smallest positive x at which $A''(x)$

changes sign occurs when $x = \tan x$ ie $x \approx 4.493$.

$$y \approx A(4.493) \approx \int_0^{4.493} \dfrac{\sin t}{t}\, dt \approx 1.656.$$ So the approximate coordinates of the

inflection point are $(4.493, 1.656)$.

10. a) $0,\ 3,\ 7\dfrac{1}{2},\ 5\dfrac{1}{2}$

b) -3 \qquad c) 3

d) $0 < x < 5$

Worksheet 6 – Calculators are allowed

1. The approximate area of the region enclosed by the graphs of $y = 4 - 3e^{-x}$ and

   $y = e^x$ is

   (A) 0.723   (B) 1.16   (C) 0.394   (D) 0.927   (E) 0.481

2. The rate at which ice is melting in a pond is given by $\dfrac{dV}{dt} = \sqrt{1 + \left(\ln(t+1)\right)^2}$

   where V is the volume of ice in cubic feet and t is the time in minutes. What

   amount of ice has melted in the first 4 minutes?

   (A) 6.74 ft^3 (B) 5.85 ft^3 (C) 4.91 ft^3 (D) 5.23 ft^3 (E) 4.72 ft^3

3. Consider the function F defined so that $F(x) - 4 = \int_1^x \cos^2\left(\dfrac{\pi t}{4}\right) dt$. The value

   of $F(2) + F'(3)$ is

   (A) 3.15   (B) 3.27   (C) 3.39   (D) 3.71   (E) 4.68

4. If $\int_2^4 f(x)\, dx = 5$ and $\int_2^7 f(x)\, dx = 11$ then $\int_4^7 \left[f(x) + 3\right] dx$

   (A) 6   (B) 9   (C) 12   (D) 15   (E) 18

5. Which of the following is a true statement?

   (A)   If f in increasing on the interval [a,b], then the minimum value of f

        on [a,b] is $f(a)$.

   (B)   The value of $\int_a^b f(x)\, dx$ must be positive.

   (C)   If $\int_a^b f(x)\, dx > 0$ then f is nonnegative for all x in [a,b].

(D) $\displaystyle\int_a^b \left[f(x)\cdot g(x) \right] dx = \int_a^b f(x)\, dx \cdot \int_a^b g(x)\, dx$

(E) None of these.

(A) A   (B) B   (C) C   (D) D   (E) E

6. Suppose $F(x) = \displaystyle\int_0^{x^2} \frac{1}{2+t^3}\, dt$ for all real x; then $F'(-1) =$

(A) 2   (B) 1   (C) $\dfrac{1}{3}$ (D) -2   (E) $-\dfrac{2}{3}$

7. If f and g are continuous functions such that $g'(x) = f(x)$ for all x, then

$$\int_2^3 f(x)\, dx =$$

(A) $g'(2) - g'(3)$ (B) $g'(3) - g'(2)$ (C) $g(3) - g(2)$

(D) $f(3) - f(2)$ (E) $f'(3) - f'(2)$

8. Let $G(x) = \displaystyle\int_{-3}^{x} g(t)\, dt,\ -3 \le x \le 3$, where g is

the function graphed in the figure.

The value of $G(3)$ is

(A) 5.5   (B) 4.5

(C) 3   (D) 2

(E) None of these

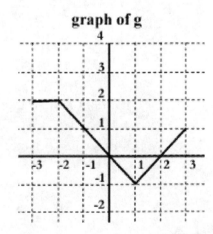

graph of g

9. Let $f(x) = \begin{cases} 3 & \text{if } x < 0 \\ 2x+3 & \text{if } x \geq 0 \end{cases}$ and let $F(x) = \int_{-2}^{x} f(t)\,dt$.

Which of the following statements is true?

      I. $F(1) = 10$

      II. $F'(1) = 5$

      III. $F''(1) = 2$

  (A) I only   (B) II only   (C) I and II only

  (D) II and III only   (E) I, II, III

10. If $\int_{1}^{5}\left[2f(x)+3\right]\,dx = 20$, then $\int_{1}^{5} f(x)\,dx =$

  (A) 1   (B) 2   (C) 3   (D) 4   (E) None of these

11. Let $f(x) = x^2 - x^3$. The area of the region between the x-axis and the graph

of f is:

  (A) $\dfrac{1}{3}$ (B) $\dfrac{1}{4}$ (C) $\dfrac{1}{6}$ (D) $\dfrac{1}{12}$ (E) none of these

12. The area of the region enclosed by the graphs of $y = kx$ and $y = x^2$, where

k is a positive constant is:

  (A) $\dfrac{5k^2}{6}$ (B) $\dfrac{k^2}{6}$ (C) $\dfrac{k^3}{6}$ (D) $\dfrac{5k^3}{6}$ (E) none of these

13. $\displaystyle\int_{-1}^{\ln 2} 4e^{-x}\,dx$

  (A) 10   (B) $4e - 2$ (C) 12   (D) 8.85   (E) 6.28

14. Suppose f is a continuous function such that $\int_0^3 f(x)\,dx = -2$. Which of the following statements is true?

    I.   If f is an even function then $\int_{-3}^3 f(x)\,dx = -4$.

    II.   If f is an odd function then $\int_{-3}^3 f(x)\,dx = 0$.

    III. $\int_2^5 f(x-2)\,dx = -2$.

(A) I only   (B) II only   (C) I and II only

(D) II and II only   (E) I, II, III

15. If $\int_{-3}^3 \left(x^{15} + k\right)\,dx = 24$, then $k =$

(A) -12   (B) 12   (C) -4   (D) 4   (E) none of these

16. The position of a particle moving along a straight line is given by

    $s(t) = \int_0^t \dfrac{1}{u^2 + 1}\,du$. The velocity at $t = 1$ is

(A) $-\dfrac{1}{2}$ (B) $-\dfrac{1}{4}$ (C) 0   (D) $\dfrac{1}{2}$ (E) $\dfrac{1}{4}$

17. The figure shows the graph of f.

Determine a so that $\displaystyle\int_{-3}^{a} f(t)\,dt$ will be as

small as possible.

$a =$

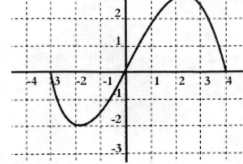

(A) -3   (B) -2   (C) 0   (D) 4   (E) none of these

18. Suppose that f is a continuous function defined on the interval $[2,5]$ and

that $\displaystyle\int_{2}^{5} f(x)\,dx = 3$. Then which of the following is true?

I. $\displaystyle\int_{2}^{5}\left[f(x)\right]^2 dx = 9$ II. $\displaystyle\left[\int_{2}^{5} f(x)\,dx\right]^2 = 9$ III. $\displaystyle\int_{2}^{5}\sqrt{f(x)}\,dx = \sqrt{3}$

(A) I, III only   (B) II only   (C) I, II, III   (D) I only

(E) They are all false

19. Consider the linear function f with the following properties:

i) $\displaystyle\int_{1}^{3} f(x)\,dx = 10$ ii) $\displaystyle\int_{1}^{5} f(x)\,dx = 28$

If $f(x) = ax + b$, then $a + b =$

(A) 1   (B) 2   (C) -1   (D) 0   (E) 3

20. $F(x) = \displaystyle\int_{\frac{\pi}{4}}^{x} \sin^2 t\,dt$. $F(x)$ has an inflection point when $x =$

(A) $\dfrac{\pi}{2}$ (B) $\dfrac{\pi}{6}$ (C) $\dfrac{\pi}{4}$ (D) $\dfrac{\pi}{3}$ (E) none of these

Answers to Worksheet 6

1. C	8. E	15. D
2. B	9. E	16. D
3. E	10. D	17. C
4. D	11. D	18. B
5. A	12. C	19. E
6. E	13. B	20. A
7. C	14. E	

Riemann Sums

To find the area under a curve whose equation is not capable of being integrated, except by graphing calculator, we can use an approximation technique known as Riemann Sums – named after the 19th century mathematician Georg Riemann.

For example, suppose we wished to find the area of the region R bounded by $x=3$, $x=15$, $f(x)=\sqrt{16-x}\ln x$, and the x axis.

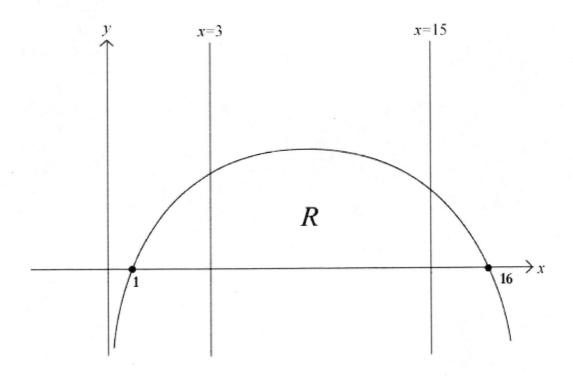

To find the area by hand methods, i.e. without a calculator, is impractical but an approximation method is available as follows.

Let us split up the region R into 6 sections formed by drawing vertical lines at $x = 5$, $x = 7$, $x = 9$, $x = 11$, $x = 13$, and $x = 15$ as shown below. We wish to add up the 6 "rounded-topped" sections to find the area of region R.

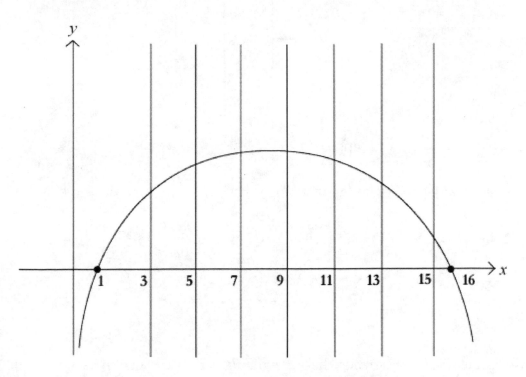

There are many ways to approximate the areas of these rounded-topped sections.

1) We can use rectangles whose height is the left-hand edge of each section.

i.e.

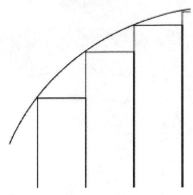

in which case the area is approximately

$$= 2 \text{ times} \left[f(3) + f(5) + f(7) + f(9) + f(11) + f(13) \right]$$

$$= 2 \text{ times} \left[\sqrt{13} \ln 3 + \sqrt{11} \ln 5 + \sqrt{9} \ln 7 + \sqrt{7} \ln 9 + \sqrt{5} \ln 11 + \sqrt{3} \ln 13 \right]$$

$$= 61.509$$

2) We can use rectangles whose height is the right-handed edge of each section.

i.e.

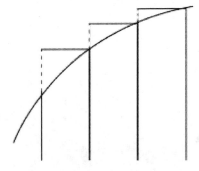

in which case the area is approximately

$$= 2 \left[f(5) + f(7) + f(9) + f(11) + f(13) + f(15) \right]$$

$$= 2 \left[\sqrt{11} \ln 5 + \sqrt{9} \ln 7 + \sqrt{7} \ln 9 + \sqrt{5} \ln 11 + \sqrt{3} \ln 13 + \sqrt{1} \ln 15 \right] = 59.003$$

3) We can use rectangles whose heights are the y values of the function at the mid-point of each section.

i.e.

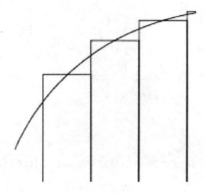

in which case the area is approximately

$$= 2\left[f(4) + f(6) + f(8) + f(10) + f(14) \right]$$

$$= 2\left[\sqrt{12}\ln 4 + \sqrt{10}\ln 6 + \sqrt{8}\ln 8 + \sqrt{6}\ln 10 + \sqrt{4}\ln 12 + \sqrt{2}\ln 14 \right]$$

$$= 61.384$$

4) We can use trapezoids formed by joining top points of each section.

i.e.

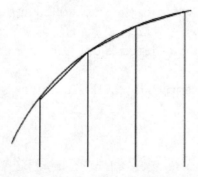

in which case the area is approximately

$$= 2\left[\frac{f(3)+f(5)}{2} + \frac{f(5)+f(7)}{2} + \frac{f(9)+f(11)}{2} + \frac{f(11)+f(13)}{2} + \frac{f(13)+f(15)}{2}\right]$$

$$= 2\text{times}\frac{1}{2}\left[f(3)+2f(5)+2f(7)+2f(9)+2f(11)+2f(13)+f(15)\right]$$

$$= \left[\sqrt{13}\ln 3 + 2\sqrt{11}\ln 5 + 2\sqrt{9}\ln 7 + 2\sqrt{7}\ln 9 + 2\sqrt{5}\ln 11 + 2\sqrt{3}\ln 13 + \sqrt{1}\ln 15\right]$$

$$= 60.256$$

In fact, using a graphing calculator we can ascertain that the area of region R is

61.013.

This shows that in the four cases, the four approximations are quite good with the

mid-point approximation being the best in this example.

Of course the more sections into which we split up the region R the greater the

accuracy.

The accuracy and the fact of whether the trapezoidal approximation is an

overestimate or an underestimate depends upon the concavity of the curve. For

example, the trapezoidal rule approximation is an underestimate because the curve is

concave down for $3 \le x \le 15$.

For the Riemann Sums using rectangles, attention must be paid to the slope of the curve

to gauge whether it is an underestimate or an overestimate.

Worksheet 7

1. The graph of the function f over the interval $[1,7]$ is shown below. Using values from the graph, find the Trapezoid rule estimates from the integral

   $\int_{1}^{7} f(x)\,dx$ using the indicated

   number of subintervals.

   a) $n = 3$

   b) $n = 6$

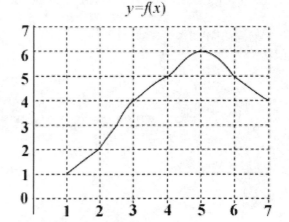

$y = f(x)$

2. Use (a) the Trapezoidal Rule and (b) the Midpoint rule with 4 equal subdivisions to approximate the definite integral

   $$\int_{-1}^{3} |2x - 3|\,dx\,.$$

3. Estimate $\int_{0}^{1} \cos(x^3)\,dx$ using (a) the Trapezoidal Rule and (b) the Midpoint rule with $n = 4$. From a graph of the integrand, decide whether your answer to the Trapezoidal Rule is an underestimate or overestimate.

4. The graph of g is shown in the figure. Estimates of $\displaystyle\int_0^2 g(x)\,dx$ were computed using the left, right, trapezoid and midpoint rules, each with the same number of subintervals. The answers recorded were 1.562, 1.726, 1.735, and 1.908.

a) Match each approximation with the corresponding rule.

b) Between which two approximations does the true value of the integral lie?

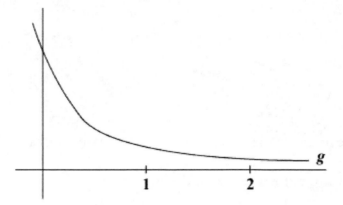

5. The graph of f over the interval [1,9] is shown in the figure. Using the data in the figure, find a midpoint approximation with 4 equal subdivisions for

$$\int_1^9 f(x)\,dx.$$

(A) 20   (B) 21   (C) 22   (D) 23   (E) 24

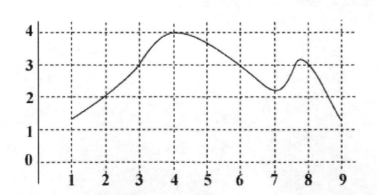

6. Let R be the region in the first quadrant bounded by the graph of

$f(x) = \sqrt{1 + e^{-x}}$, the line $x = 2$, and the x-axis.

(a) Write an integral that gives the area of R.

(b) Use the Trapezoidal Rule with $n = 4$ to approximate the area. You must

show the numbers that lead to your answer.

7. Answer the following questions about the function f, whose graph is shown

below.

(a) Find $\lim_{x \to 0} f(x)$.

(b) Find $\lim_{h \to 0^-} \dfrac{f(0+h) - f(0)}{h}$.

(c) Find $\lim_{x \to 0} f'(x)$.

(d) Find $\displaystyle\int_{-1}^{0} f(x)\,dx$.

(e) Approximate $\displaystyle\int_{-2}^{4} f(x)\,dx$ using the Trapezoidal Rule with $n = 3$

subdivisions.

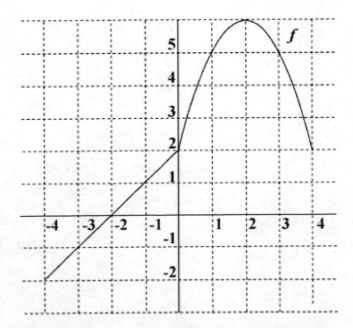

8. The Mean Value Theorem guarantees the existence of a special point on the graph of $y = \sqrt{x}$ between $(1,1)$ and $(9,3)$. What are the co-ordinates of this point?

(A) $(1,1)$

(B) $(2, \sqrt{2})$

(C) $(3, \sqrt{3})$

(D) $(4,2)$

(E) None of the above

9. Let R be the first quadrant region bounded by the graph of $f(x) = \sqrt{1 + x^2}$ from $x = 1$ to $x = 4$. Use the Trapezoidal Rule with 3 equal subdivisions to approximate the area of the region R.

(A) 8.15   (B) 8.17   (C) 8.19   (D) 8.21   (E) 8.23

10. The following table lists the known values of a function f.

x	1	2	3	4	5
$f(x)$	0	1.1	1.4	1.2	1.5

If the Trapezoidal Rule is used to approximate $\int_{1}^{5} f(x)\,dx$ the result is

(A) 4.39   (B) 4.42   (C) 4.45   (D) 4.48   (E) 4.51

11. A cedar log 6 m long is cut at 1 m intervals and its cross-sectional areas A (at a distance x from the end of the log) are recorded in the following table.

x (m)	0	1	2	3	4	5	6
A (m²)	0.64	0.62	0.6	0.56	0.52	0.46	0.38

Find an approximate volume of the log in cubic metres.

Answers to Worksheet 7

1. a) 25   b) 24.5

2. a) 9   b) 8

3. a) 0.919 underestimate   b) 0.939

4. a) Right – 1.562   Mid-point – 1.726   Trapezoid – 1.735   Left – 1.908

    b) between 1.726 and 1.735

5. E

6. a) $\int_0^2 \sqrt{1+e^{-x}}\,dx$ b) 2.72 (actual value is 2.39)

7. a) 2   b) 1   c) does not exist   d) 1.5   e) 18

8. D

9. B

10. C

11. 3.27 cubic metres.

CHAPTER 14

Position, Velocity, Acceleration II

Earlier, in Chapter 7, we learned that $a = \dfrac{dv}{dt}$ where a is acceleration, v is velocity

and t is time.

When a stone is thrown into the air, its velocity depends upon acceleration due to

gravity and its initial velocity. We know that acceleration due to gravity is about

(9.81 m/sec) per second but let us assume for simplicity's sake that this acceleration is

(9.8 m/sec) per second.

We consider <u>upwards</u> motion as positive and hence

$$a = -9.8$$

i.e. $\quad \dfrac{dv}{dt} = -9.8$

By 'going backwards' we can deduce that $v = -9.8t + c$ where c is some constant.

Let the initial velocity be v_0 and then it is clear that $c = v_0$ because when $t = 0$, $v = v_0$.

i.e. $\quad v = -9.8t + v_0$

If the initial velocity were (say) 100 m/sec then

$$v = -9.8t + 100$$

and we can easily find the velocity at any given time t.

Incidentally since 9.8 is measured in (m/sec) per second then v is measured in metres/second and t is measured in seconds.

Similarly we can go backwards from v to position x, since $v = \dfrac{dx}{dt}$

$$v = -9.8t + v_0$$

$$\therefore \quad \frac{dx}{dt} = -9.8t + v_0$$

$$\therefore \quad x = -\frac{9.8t^2}{2} + v_0 t + k \ ,$$

where k is a constant representing the initial position.

Remember that this last equation refers to the situation for motion subject to gravity only and is derived from Newton.

Example

Question: A particle is projected vertically upwards from the ground with an initial velocity of 50 metres per second. Find the maximum height which the particle attains.

Answer:
$$a = -9.8 \qquad (1)$$

$$\therefore \quad v = -9.8t + c$$

But $v_0 = 50$, $\quad \therefore \ c = 50$

$$\therefore \quad v = -9.8t + 50 \qquad (2)$$

When the particle attains its maximum height its velocity is zero and hence at this instant

$$0 = -9.8t + 50$$

i.e. $t = 5.102$.

From (2) we can go backwards again to deduce that

$$x = -4.9t^2 + 50t + k \quad \text{where } k \text{ is a constant representing the}$$

initial position

When $t = 0$, $x = 0$ (particle is projected from the ground) and hence $k = 0$.

$$\therefore \quad x = -4.9t^2 + 50t \qquad (3)$$

When $t = 5.102$, $x = 127.55$

$\therefore$ The maximum height attained by the particle is 127.55 metres.

Example

Question: At time $t = 0$ a runner is running at a velocity of 400 metres per minute. The runner starts to tire and slows down at a rate directly proportional to time.

i.e. $a(t) = kt$ where k is some negative constant. The runner comes to a halt after 10 minutes. Assume the runner's initial position is zero.

a) Find the velocity of the runner at any time t $(0 \leq t \leq 10)$.

b) Find the velocity after 5 minutes.

c) Find the total distance travelled by the runner.

Answer:　a)　　$a = kt$

$$\therefore \quad v = \frac{k}{2}t^2 + 400 \quad \text{(since the initial velocity is 400)}$$

When $t = 10$, $v = 0$, $\therefore$ $k = -8$.

$$\therefore \quad v = -4t^2 + 400 \quad \circledast$$

b)　　$v(5) = 300$

c)　　From $\circledast$ we can deduce that

$$x = -\frac{4}{3}t^3 + 400t.$$

Note that the constant is zero since $x(0) = 0$.

The total distance travelled is $x(10)$ which is 2667 metres

(approx.) Note that since the runner does not change direction

then $x(10)$ does represent the distance travelled as well as the

position of the runner.

Example

Question:　A particle moves along a horizontal line such that its acceleration at

any time t seconds $(t \geq 0)$ is given by

$$a(t) = \pi \cos\left(\frac{\pi t}{2}\right)$$

Initially, i.e. at time $t = 0$, the velocity is 1 and its position after 1

second is 0 i.e. $v(0) = 1$ and $x(1) = 0$.

a) Find an expression for the velocity $v(t)$ of the particle at any time t.

b) Find an expression for the position $x(t)$ of the particle at any time t.

c) Find when the particle is at rest in the first 3 seconds.

d) Find the total distance travelled in the first 4 seconds.

e) Find the values of t for which the particle's <u>speed</u> is increasing in the interval $0 \leq t \leq 4$.

Answer:

a) $a(t) = \pi \cos \dfrac{\pi t}{2}$

$\therefore \quad v(t) = 2\sin\dfrac{\pi t}{2} + k \quad$ where k is some constant

But $v(0) = 1 \quad \therefore \quad 2\sin\dfrac{\pi}{2} \cdot 0 + k = 1$

i.e. $\quad k = 1$

$\therefore \quad v(t) = 2\sin\dfrac{\pi t}{2} + 1$

b) $v(t) = 2\sin\dfrac{\pi}{2}t + 1$

$\therefore \quad x(t) = -\dfrac{4}{\pi}\cos\dfrac{\pi}{2}t + t + c$ where c is some constant.

$x(1) = 0 \qquad \therefore \quad -\dfrac{4}{\pi}\cos\dfrac{\pi}{2} \cdot 1 + 1 + c = 0$

$\therefore \qquad c = -1$

$$\therefore \quad x(t) = -\frac{4}{\pi}\cos\frac{\pi}{2}t + t - 1$$

---

c)   When the particle is at rest $v = 0$

$$\therefore \quad 2\sin\frac{\pi}{2}t + 1 = 0$$

i.e. $$\sin\left(\frac{\pi}{2}t\right) = -\frac{1}{2}$$

i.e. $$\frac{\pi}{2}t = \frac{7\pi}{6}, \frac{11\pi}{6}\ldots$$

i.e. $$t = \frac{7}{3}, \frac{11}{3}\ldots$$

We want the time(s) in the first 3 seconds.

$$\therefore \quad t = \frac{7}{3} \text{ seconds}$$

d)   In the first 4 seconds the motion of the particle is:

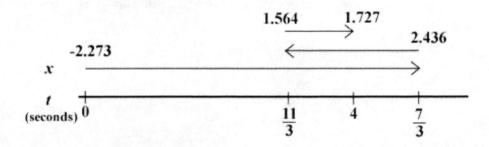

The total distance is hence:

$$\left|x\left(\frac{7}{3}\right) - x(0)\right| + \left|x\left(\frac{7}{3}\right) - x\left(\frac{11}{3}\right)\right| + \left|x(4) - x\left(\frac{11}{3}\right)\right| = 5.744 \text{ (approx)}.$$

These values were all obtained using a graphing calculator. A

shorter method, which will be explained later is to evaluate

$$\int_0^4 |v(t)|\, dt \ .$$

In general, the total distance travelled by a particle from time

$$t = a \ \text{to} \ t = b \ \text{is} \ \int_a^b |v(t)|\, dt \ .$$

e) As shown in Chapter 7, speed is increasing occurs when velocity is positive and acceleration is positive or when velocity is negative and acceleration is negative i.e. speed is increasing when velocity and acceleration have the same sign.

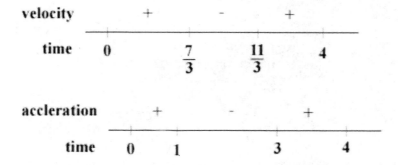

The times when v and t have the same sign are

$$0 \le t \le 1 \quad \text{OR} \quad \frac{7}{3} \le t \le 3 \quad \text{OR} \quad \frac{11}{3} \le t \le 4$$

And hence are the times when speed is increasing.

Worksheet 1

INTEGRATION - LINEAR MOTION

1. The acceleration of a rocket is given by the formula $a = 64t(20 - t)$ where

   $0 \le t \le 20$. If the rocket starts with a velocity of 200 find:

   a) the velocity at any time t.

   b) the velocity when $t = 20$.

   c) the position at any time t.

2. A particle accelerates at a rate of a (feet per second) per second after t

   seconds where $a = 2t^2 + 4t$. After three seconds the particle is travelling with

   a velocity of 40 feet per second. Find its initial velocity and position after 2

   seconds.

3. A particle starts at the origin and moves along the x axis so that its velocity at

   the end of t seconds is given by $v = 1 - 2\cos t$ where $0 \le t \le 4\pi$. Find the

   position and acceleration at any time t, the times when the particle is

   stationary, maximum and minimum velocity, and maximum acceleration.

4. Acceleration due to gravity is 9.8 metres per second per second. Find the

   initial velocity of an object thrown vertically upwards from the ground so that

   it attains its maximum height after 6 seconds. Find also the maximum height.

5. A particle moves such that its acceleration from a starting point O is given by

   $a = -4t$. If it starts with an initial velocity of 4 metres per second when does it

   return to its starting position?

6. The acceleration of a rocket after t seconds is given by the formula

$a = 15\sqrt{t+100} - 10$ metres per second per second. If the initial position and velocity are both zero find:

a) the position and velocity after t seconds.

b) velocity after 5 seconds.

Answers to Worksheet 1

1. a) $640t^2 - \dfrac{64}{3}t^3 + 200$ b) $v_{20} = 85533\dfrac{1}{3}$ c) $s = \dfrac{640}{3}t^3 - \dfrac{16}{3}t^4 + 200t$

2. 4 feet per second. 16 feet.

3. $s = t - 2\sin t$, $a = 2\sin t$

   particle is stationary at times $\dfrac{\pi}{3}, \dfrac{5\pi}{3}, \dfrac{7\pi}{3}, \dfrac{11\pi}{3}$

   maximum velocity is 3, minimum velocity is -1, maximum acceleration is 2

4. 58.8 metres per second. Maximum height is 176.4 ft.

5. $\sqrt{6}$ seconds

6. a) $s = 4(t+100)^{\frac{5}{2}} - 5t^2 - 10000t - 400000$, $v = 10(t+100)^{\frac{3}{2}} - 10t - 10000$

   b) 709.3 ft/sec

Worksheet 2

INTEGRATION - LINEAR MOTION

1. A parachutist jumps from an aircraft flying horizontally 4885 feet above the ground. During the first 10 seconds he falls at a velocity v where $v = -32t$ feet per second where t is the time in seconds from the instant when he jumped. After 10 seconds his parachute opens and then his downward velocity is $-20 - \dfrac{30000}{t^3}$ feet per second where t is time in seconds from the instant when the parachutist jumped from the plane.

    a) At what height does his parachute open?

    b) When does the parachutist land?

2. A man can jump 4 feet into the air on the surface of the Earth where g is 32 feet per second per second. How high could he jump on

    a) Mars (where g is 8 feet per second per second)?

    b) The Moon (where g is 5 feet per second per second)?

    b) Jupiter (where g is 80 feet per second per second)?

3. The acceleration of a particle is given by $a = \sqrt{t+9}$ feet per second per second. It starts with an initial velocity of 5 feet per second from an initial position of 65 feet.

    a) Find v after 7 seconds.

    b) Find position after 7 seconds.

4. A car entering a speed zone is observed to have a speed of 60 feet per second at that instant. The car brakes at a steady deceleration of 5 feet per second per second.

   a) Find the speed of the car after t seconds.

   b) The distance travelled in the first 5 seconds.

   c) Find how long it takes for the car to travel 200 feet.

   d) Find the distance travelled when $v = 30$ feet per second.

5. True or false? (Assume $c > b > a > 0$)

$$\int_a^b f(x)\,dx + \int_b^c f(x)\,dx = \int_a^c f(x)\,dx$$

Answers to Worksheet 2

1. a) 3285 feet b) 166.78 seconds

2. a) 16 feet   b) 25.6 feet c) 1.6 feet

3. a) 29.66 feet b) 182.27 feet

4. a) $v = 60 - 5t$ b) $237\frac{1}{2}$ feet   c) 4 seconds   d) 270 feet

5. True

Question: A car travels on a highway with velocity given by $v(t) = 3t^2 - 27t + 60$ for $t > 0$.

a) Write a definite integral to denote the distance travelled by the car between $t = 0$ and $t = 6$.

b) Evaluate the integral in part a).

c) Find a function $x(t)$ representing the position of the car if $x(0) = 0$ (i.e. it starts at the zero position).

d) Find the times when speed is increasing.

Answer: a) The car stops when $v(t) = 0$.

i.e. $3t^2 - 27t + 60 = 0$

$3(t-4)(t-5) = 0$

i.e. when $t = 4$ or $t = 5$.

i.e. the car stops when $t = 4$ and when $t = 5$.

Note that when $4 \le t \le 5$, the car is going backwards, since velocity is negative, and hence to find the <u>distance</u> travelled from $t = 4$ to $t = 5$ we need to evaluate the integral of velocity when $t = 4$ and subtract the integral of velocity when $t = 5$.

i.e. We need to evaluate $\int_5^4 v(t)\, dt$.

This can be written $\int_4^5 |v(t)|\, dt$.

When $0 < t < 4$ or $t > 5$ the car is going forwards and hence there is no problem since $|v(t)| = v(t)$ for those values of t.

It follows that the total distance travelled for $0 \le t \le 6 = \int_0^6 |v(t)| \, dt$.

This can be evaluated easily using a graphing calculator but by hand it is more difficult and perhaps a more practical approach would be to say that the total distance travelled is

$$\int_0^6 v(t) \, dt - \int_4^5 v(t) \, dt + \int_5^6 v(t) \, dt.$$

b) By calculator $\int_0^6 |v(t)| \, dt = \int_0^6 |3t^2 - 27t + 60| \, dt = 91$ (approx).

By hand,

$$\int_0^4 3t^2 - 27t + 60 \, dt - \int_4^5 3t^2 - 27t + 60 \, dt + \int_5^6 3t^2 - 27t + 60 \, dt$$

$$= \left[t^3 - \frac{27}{2}t^2 + 60t \right]_0^4 - \left[t^3 - \frac{27}{2}t^2 + 60t \right]_4^5 + \left[t^3 - \frac{27}{2}t^2 + 60t \right]_5^6$$

$$= \left[(64 - 216 + 240) - (0) \right] - \left[(125 - 337.5 + 300) - (64 - 216 + 240) \right]$$

$$+ \left[(216 - 486 + 360) - (125 - 337.5 + 300) \right]$$

$$= 88 - (-0.5) + 2.5$$

$$\underline{\underline{= 91}}$$

Worksheet 3

1. A particle with a velocity at any time v given by $v(t) = 2e^{2t}$ moves in a straight line. How far does the particle travel during the time when its velocity increases from 2 to 4?

   (A) 1   (B) 2   (C) 3   (D) e^4 (E) $e^8 - e^4$

2. If the position s of a point on a number line is given by the formula

   $$s = \frac{t^2 - 6t + 5}{(t+1)^2}, \ t \geq 0 \ (t \text{ is in seconds, } s \text{ is in metres), find:}$$

   a) the average velocity over the first 3 seconds of motion.

   b) the total distance travelled by the point in the first 3 seconds of motion.

3. If the acceleration of a particle on a horizontal line is x where x represents the position in metres from the starting position of zero and if the initial velocity is 8 centimetres per second find the velocity of the particle when it is in a position of 15 centimetres.

4. If the acceleration of a particle on a horizontal line is given by $a(t) = \sqrt{t + 4}$ and if the particle starts from rest find the velocity after 12 seconds.

5. The figure shows the velocity of two marathon runners racing against each other. They start together at the same time on the same course.

a) Which runner is ahead after the first minute? Explain your answer.

b) Which runner is ahead after the second minute? Explain your answer.

c) When are the runners about level?

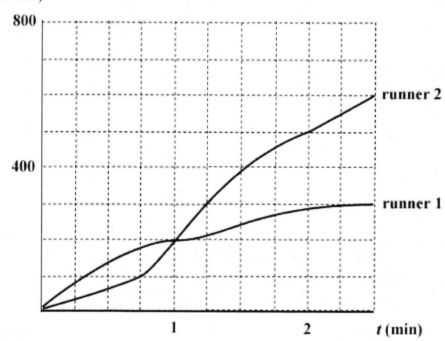

6. A particle moves along the x axis so that its position at any time $0 \le t \le 4$ is given by $x(t) = t^4 - 8t^3 + 18t^2 - 9t + 5$. For which of the following values of t is the <u>speed</u> the greatest?

(A) $t = 1$ (B) $t = 1.5$ (C) $t = 2$

(D) $t = 3$ (E) $t = 3.5$

7. The velocity of a baseball hit along by the ground by Vernon Wells is given by

the formula $v = 20\sqrt{4 - \dfrac{s}{100}}$ where s represents the distance the ball has

travelled along the ground in feet and v is in feet per second. Find:

a) The velocity of the ball on impact.

b) The distance the ball travels before it stops rolling.

c) Prove that the acceleration of the ball is a constant. Find this constant.

8. A car is moving forward and backward along a straight road from A to B,

starting from A at time $t = 0$. The car's velocity is given by

$v(t) = 1 + 2\sin\left(\dfrac{\pi t}{6}\right)$ where t is in minutes and v is in km/min. The graph of

the velocity function is given below.

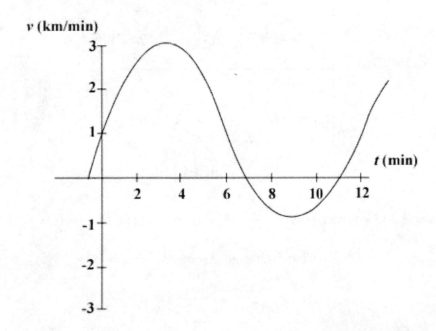

a) What is the velocity of the car at $t = 0$?

b) At what time(s) t does the car change direction?

c) Find the average velocity of the car between $t = 0$ and $t = 9$.

9. A particle moves along a line so that at any time $t \geq 0$ its velocity is given by

$v(t) = \dfrac{t}{1+t^2}$. At time $t = 0$ the position of the particle is $s(0) = 5$.

a) Determine the maximum velocity of the particle. Justify your answer.

b) Determine the position of the particle at $t = 3$.

c) What is the total distance travelled by the particle from $t = 1$ to $t = 3$.

d) Find the limiting value of the velocity as t increases without bound.

10. A car is moving along a straight road from A to B, starting from A at time $t = 0$. Below is a graph of the car's velocity (positive direction from A to B), plotted against time.

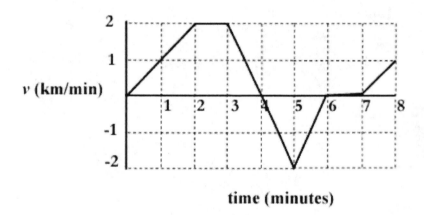

a) How many kilometres away from A is the car at time $t = 6$?

b) At what time does the car change direction? Explain briefly.

Answers to Worksheet 3

1. A

2. a) -1.75   b) 5.42

3. 17

4. $\dfrac{112}{3}$

5. a) runner 1 because the area under velocity curve denotes distanced travelled.

    b) runner 2.

    c) $1\dfrac{1}{2}$ minutes (approx.)

6. D

7. a) 40   b) 400   c) -2

8. a) 1   b) $t=7$ and $t=11$ c) $1+\dfrac{4}{3\pi}$

9. a) $\dfrac{1}{2}$ b) $\dfrac{1}{2}\ln 10+5$ c) $\dfrac{1}{2}\ln 5$ d) 0

10. a) 3   b) At $t=4$ and $t=7$.

CHAPTER 15

Differential Equations

In many natural conditions the rate at which the amount of an object changes is directly proportional to the amount of the object itself. For example:

1) The marginal cost of a product varies directly with the number of units manufactured.

2) The rate at which certain populations increase varies directly with the size of the population, i.e. $\dfrac{dP}{dt} = kt$ where $P(t)$ is the population at time t.

3) The temperature of a body changes at a rate proportional to the positive difference between the temperature of the body and its immediate environment. This is often called Newton's Law of Temperature Change. In the case of a body cooling we have $\dfrac{dT}{dt} = k(T - E)$. Where E is the fixed temperature of the environment and $T(t)$ is the temperature of a body at time t. k is a positive constant. In the case of a body heating, e.g. putting an egg into a saucepan of boiling water, we have $\dfrac{dT}{dt} = k(E - T)$ where k is positive.

4) If a substance is poured into a glass of water and the substance dissolves then the rate at which the substance dissolves is directly proportional to the quantity of the substance left undissolved.

e.g. $\dfrac{dS}{dt} = kS$ where S is the quantity of substance undissolved.

Here k is a negative constant.

5) A radioactive substance decays at a rate directly proportional to the amount of substance. $\dfrac{dM}{dt} = kM$ where M is the amount of substance present. k again is a negative constant.

Each of the foregoing examples is modeled by the differential equation $\dfrac{dy}{dt} = ky$. To solve a differential equation expressed in this form we mean find y explicitly in terms of t. This can be effected as follows:

$$\frac{dy}{dt} = ky$$

$\therefore$ $$\frac{1}{y}\frac{dy}{dt} = k$$

Integrating both sides of the equation with respect to t yields

$$\ln|y| = kt + c$$

For most practical problems y is always positive and so we can say $\ln y = kt + c$ for some constant c.

i.e. $\qquad y = e^{kt+c}$

$$= e^c \cdot e^{kt}$$

i.e. $\qquad y = Ae^{kt}$ where A is a constant.

Note that this equation is a solution to the original equation because

if $\qquad y = Ae^{kt}$

then $\qquad \dfrac{dy}{dt} = Ae^{kt} \cdot k = ky$ as required.

Example 1

Question: $\qquad$ Suppose $\dfrac{dy}{dx} = 4xy$ and $y = 1$ when $x = 2$. Find y when $x = 3$.

Answer: $\qquad \dfrac{dy}{dx} = 4xy$

$\therefore \qquad \dfrac{1}{y}\dfrac{dy}{dx} = 4x$

$\qquad \ln y = 2x^2 + c$

When $x = 2$, $y = 1$ (given)

$\therefore \qquad \ln 1 = 8 + c$

$\qquad 0 = 8 + c$

$\qquad -8 = c$

$\therefore \qquad \ln y = 2x^2 - 8$

i.e. $\qquad y = e^{2x^2 - 8}$

$\therefore \qquad$ When $x = 3$, $y = e^{10}$.

Example 2

Question: $\dfrac{dy}{dx} = \dfrac{2x}{e^y}$ and $y = 0$ when $x = 1$.

Find y when $x = 2$.

Answer: $\dfrac{dy}{dx} = \dfrac{2x}{e^y}$

$\therefore \qquad e^y \dfrac{dy}{dx} = 2x$

Integrating both sides with respect to x we get:

$$e^y = x^2 + c$$

When $x = 1$, $y = 0$ (given)

$\therefore \qquad e^0 = 1^2 + c$

i.e. $0 = c$

i.e. $e^y = x^2$

When $x^2 = 2$, $e^y = 4$

$\therefore \qquad y = \ln 4$.

As a precautionary note, remember that the process of differentiation is not a $1 - 1$ function since, for example, $D\!\left(x^2\right) = D\!\left(x^2 + 1\right)$.

It follows that integration, the inverse operation of differentiation, is <u>NOT</u> a function since for example $\int 2x\,dx = x^2 + k$ for <u>any</u> constant k.

Therefore to solve differential equations it is necessary to be given some initial values of the variables.

Example 3

Question: The rate at which the population of a bacteria culture grows is proportional to the number of bacteria present. If the number of bacteria grew from 1000 to 5000 in 10 hours find the number of bacteria after 15 hours.

* Answer: Let $B(t)$ be the number of bacteria present at time t.

i.e. $$\frac{dB}{dt} = kB$$

∴ $$\frac{1}{B}\frac{dB}{dt} = k$$

∴ $\ln B = kt + c$ (note that $B > 0$ for all t)

$B = e^{kt+c}$

$B = e^c e^{kt}$

When $t = 0$, $B = 1000$, ∴ $e^c = 1000$.

i.e. $B(t) = 1000e^{kt}$

When $t = 10$, $B = 5000$

∴ $5000 = 1000e^{10k}$

∴ $5 = e^{10k}$

∴ $5^{\frac{1}{10}} = e^k$

Substituting $5^{\frac{1}{10}} = e^k$ into $B(t) = 1000e^{kt}$, we get

$$B(t) = 1000 \cdot 5^{\frac{t}{10}}$$

When $t = 15$, $B = 1000 \cdot 5^{1.5}$

$$= 11180 \text{ (approx.)}$$

$\therefore$ After 15 hours, the number of bacteria is 11,180 (approx.)

Example 4

Question: The atoms of certain radioactive elements disintegrate such that it is known that the amount of such substances changes at a rate proportional to the amount present. Polonium decays into lead such that after 100 days it has lost 30 % of its initial amount. Find the half-life of Polonium.

Answer: Half-life means the number of days after which the amount of Polonium is one half the amount it was at the beginning.

Let $P(t)$ be the amount of polonium at time t days. Let P_0 be the initial amount.

$$\frac{dP}{dt} = kP \qquad \text{(where } k \text{ is a negative constant because } P \text{ is decreasing)}$$

From the previous examples we can deduce that

$$P(t) = P_0 e^{kt}$$

After 100 days the polonium has lost 30 % of its initial mass.

Therefore after 100 days, $P = 0.7P_0$.

i.e. $\quad P(100) = 0.7P_0$

Substituting in $P(t) = P_0 e^{kt}$, we get

$$0.7P_0 = P_0 e^{k100}$$

$\therefore \quad (0.7)^{\frac{1}{100}} = e^k$

$\therefore \quad P = P_0 (0.7)^{\frac{t}{100}}$ is the equation expressing the amount of

polonium present after t days.

To find its half-life we must note that, at that time, $P = \dfrac{1}{2}P_0$.

i.e. $\quad 0.5P_0 = P_0 (0.7)^{\frac{t}{100}}$

i.e. $\quad 0.5 = (0.7)^{\frac{t}{100}}$

$\therefore \quad \ln(0.5) = \dfrac{t}{100}\ln(0.7)$

i.e. $\quad \dfrac{100\ln(0.5)}{\ln(0.7)} = t$

i.e. $\quad t = 194$ (approx.)

$\therefore \quad$ The half-life of polonium is 194 days approximately.

Example 5

Question: A saucepan of boiling water cools according to Newton's Law of Temperature Change such that it cools from 100 °C to 80 °C in 5 minutes when the ambient temperature is 25 °C. How long will it take for the water to cool to 50 °C?

Answer: Let $T(t)$ be the temperature of the water after t minutes from the instant when it started to cool from 100 °C.

Then $\dfrac{dT}{dt} = k(T - 25)$

$\therefore \quad \dfrac{1}{T-25}\dfrac{dT}{dt} = k$

Integrating with respect to time we get

$\ln(T - 25) = kt + c$

$\therefore \quad T - 25 = e^{kt+c} = e^c \cdot e^{kt}$

When $t = 0$, $T = 100$

$\therefore \quad 75 = e^c e^0$

i.e. $\quad e^c = 75$

$\therefore \quad T - 25 = 75 e^{kt}$

When $t = 5$, $T = 80$

$\therefore \quad 55 = 75 e^{5k}$

$\therefore \quad \left(\dfrac{55}{75}\right)^{\frac{1}{5}} = e^k$

i.e. $\qquad e^k = \left(\dfrac{11}{15}\right)^{\frac{1}{5}}$

Substituting into $T - 25 = 75e^{kt}$ we get

$$T = 75\left(\dfrac{11}{15}\right)^{\frac{t}{5}} + 25$$

When $T = 50$ we have

$$50 = 75\left(\dfrac{11}{15}\right)^{\frac{t}{5}} + 25$$

$$\dfrac{1}{3} = \left(\dfrac{11}{15}\right)^{\frac{t}{5}}$$

Taking logarithms both sides, we have

$$\ln\left(\dfrac{1}{3}\right) = \dfrac{t}{5}\ln\left(\dfrac{11}{15}\right)$$

$$5\dfrac{\ln\left(\dfrac{1}{3}\right)}{\ln\left(\dfrac{11}{15}\right)} = t$$

$$17.71 = t$$

The water will cool to 50 °C after 17.71 minutes.

Example 6

Question: A raindrop falls with acceleration $9.81 - \dfrac{v}{3.2}$ (metres/second) per second where v is the velocity in metres/second. Find the raindrop's limiting velocity as time increases assuming the raindrop's initial velocity is 0.

Answer: acceleration $= 9.81 - \dfrac{v}{3.2}$

$\therefore \quad \dfrac{dv}{dt} = 9.81 - \dfrac{v}{3.2} = \dfrac{31.392 - v}{3.2}$

$\therefore \quad \dfrac{1}{31.392 - v} \dfrac{dv}{dt} = \dfrac{1}{3.2}$

Integrating with respect to time we get $-\ln(31.392 - v) = \dfrac{1}{3.2}t + c$

i.e. $\quad \ln(31.392 - v) = -\dfrac{t}{3.2} - c$

i.e. $\quad 31.392 - v = e^{\frac{-t}{3.2}} e^{-c}$

When $v = 0$, $t = 0$

i.e. $\quad 31.392 = e^{-c}$

i.e. $\quad 31.392 - v = 31.392 e^{\frac{-t}{3.2}}$

i.e. $\quad 31.392 - 31.392^{e^{\frac{-t}{3.2}}} = v$

As time increases without bound then $e^{\frac{-t}{3.2}}$ approaches zero.

$\therefore \quad$ The limiting velocity of the raindrop is 31.392 m/sec.

Example 7

Question: A cook monitors the temperature of a roast that is in an oven set at 180 °C. At 12 noon the temperature of the roast is 110 °C and at 1 p.m. the temperature of the roast is 160 °C. If the temperature of the roast was initially 20 °C at what time was the roast put in the oven?

Answer: Let $T(t)$ be the temperature of the roast and let t be the number of hours <u>after</u> 12 noon.

$$\frac{dT}{dt} = k(180 - T)$$

$\therefore$
$$\frac{1}{180 - T} \frac{dT}{dt} = k$$

Integrating both sides with respect to t:

$$-\ln(180 - T) = kt + c$$

$\therefore$
$$\ln(180 - T) = -kt - c$$

$\therefore$
$$180 - T = e^{-kt-c} = e^{-c}e^{-kt}$$

At 12 noon, $t = 0$ and $T = 110$

$\therefore$
$$70 = e^{-c}$$

i.e.
$$180 - T = 70e^{-kt} \qquad \text{✳}$$

At 1 p.m., $t = 1$ and $T = 160$

$\therefore$
$$20 = 70e^{-k}$$

$\therefore$
$$e^{-k} = \frac{2}{7}$$

Substituting in ✱

$$\therefore \qquad 180 - T = 70\left(\frac{2}{7}\right)^{t}$$

$$\therefore \qquad T = 180 - 70\left(\frac{2}{7}\right)^{t}$$

i.e. $\qquad 20 = 180 - 70\left(\frac{2}{7}\right)^{t}$

$$70\left(\frac{2}{7}\right)^{t} = 160$$

$$\left(\frac{2}{7}\right)^{t} = \frac{16}{7}$$

Solving by logarithms $t = -0.65988$

t is measured in hours, therefore when $t = -0.65988$, we mean -39.6

minutes (approx.)

$\therefore$ The roast was put in the oven 39.6 minutes <u>before</u> noon

i.e. at 11:20 a.m. (approx.)

Authors' Note:

Often in an example such as $\frac{dB}{dt} = kB$ the equation is re-written $\frac{1}{B}dB = kdt$ and both

sides of the equation are integrated as $\int \frac{1}{B}dB = \int kdt$.

This is called separation of variables.

While it is true that this will lead to the "correct result", it is frowned upon by the

authors because

1) Even though $\frac{dB}{dt}$ is the limit of $\frac{\Delta B}{\Delta t}$ as $\Delta t \to 0$, $\frac{dB}{dt}$ is not a fraction itself

   and should not be separated.

2) When we add the $\int$ symbol to both sides of the equation it is not

   inherently clear that this is valid because, on the face of it, we will be

   integrating the left-hand side with respect to B and integrating the right-

   hand side with respect to t .

Worksheet 1

1. If $\dfrac{dy}{dx} = y + 1$ and $y = 2$ when $x = 0$, find y when $x = 1$.

2. If $\dfrac{dy}{dx} = 2y + 3$ and $y = 2$ when $x = 0$, find y when $x = 1$.

3. If $\dfrac{dy}{dx} = 2xy^2$ and $y = 1$ when $x = 1$, find y when $x = 2$.

4. If $\dfrac{dy}{dx} = e^{y+x}$ and $y = 0$ when $x = \ln 4$, find y when $x = \ln 2$.

5. The bacteria in a certain culture increase according to the law $\dfrac{dN}{dt} = kN$ where N is the number of bacteria and k is a constant. If $N_0 = 3000$ and $N_5 = 6000$ find: a) N_1 b) t when $N = 60000$.

6. Population grows at a rate such that $\dfrac{dP}{dt} = kP$ where P is the population, k is a constant and t is measured in years. Find k so that the population doubles in 5 years.

7. Uranium disintegrates at a rate proportional to the amount present at any instant. If 20 grams are present at time zero and 16 grams are present after 10 years, how many grams will be present after 20 years? Repeat the question where the rate is proportional to the amount present <u>squared</u>.

8. The downward acceleration of a man whose fall from a practice tower is being slowed by a parachute is given by $a = 32 - 10v$.

   Show that $v = 3.2\left(1 - e^{-10t}\right)$ if his initial velocity is zero.

9. Newton's Law of Cooling states that if the temperature of an ingot t minutes after it starts to cool is T °C then $\dfrac{dT}{dt} = -k(T-15)$ where 15 °C is the room temperature. Find an expression for T in terms of t if the initial temperature is 500 °C. Find the temperature after 30 minutes if $k = 0.02$.

10. The rate at which a chemical compound dissolves is proportional to the difference between the concentration and the concentration of a saturated solution. Thus, if C is the concentration (in grams per litre) and the saturation concentration is 10, then $\dfrac{dC}{dt} = k(10-C)$. Show that $C = 10(1-e^{-kt})$. If $k = 0.01$ and t is measured in minutes, what is the concentration after one hour?

11. Radium decomposes at a rate proportional to the amount present. If 200 mg reduces to 180 mg in 100 years, how many milligrams will remain at the end of 500 years? Determine the half-life of radium.

12. A body of temperature 160 ° is immersed in a liquid of constant temperature 100 °. If it takes 2 minutes for the body to cool to 140 °, how long does it take to cool to 120 °? Assume Newton's Law of Cooling.

13. P is related to t by the formula $\dfrac{dP}{dt} = 100 + 0.05P$ where t is the time in years and $P_0 = 1000$. Find P when $t = 20$.

14. In a certain chemical reaction, the rate of conversion of a substance at time t is proportional to the quantity of the substance still untransformed at that

instant. If $\frac{1}{3}$ of the original amount of the substance has been converted

when $t = 4$ min and if an amount of 500 has been converted when $t = 8$ min,

find the original amount of the substance.

Answers to Worksheet 1

1. $3e - 1$ 2. $\frac{7e^2 - 3}{2}$ 3. $-\frac{1}{2}$ 4. $-\ln 3$

5. a) 3446   b) 21.6

6. 0.1386

7. 12.8, 13.33

9. $T = 485e^{-kt} + 15$, 281.17

10. 4.51

11. 118 mgs, 658 years

12. 5.419 minutes

13. 6155

14. 900

Worksheet 2

1. $y = f(x)$ is a function such that $\dfrac{dy}{dx} = y \cdot e^x$ and $f(0) = 1$. $f(1) =$

    (A) 15.15　　(B) 15.16　　(C) 5.57　　(D) 2.27　　(E) 3.18

2. If $\dfrac{dy}{dt} = 2y$ and if $y = 1$ when $t = 1$, what is the value of t for which $y = e^2$?

    (A) 0　　　　(B) 1　　　　(C) 2　　　　(D) $\dfrac{1}{2}$　　(E) $\dfrac{3}{2}$

3. At each point (x, y) on a certain curve, the slope of the curve is $2xy$. If the

    curve contains the point $(0,5)$ then the curve contains the point

    (A) $(1, 5e)$　　　　　　(B) $(1,1)$　　　　　　(C) $(2,10)$

    (D) $(2,25)$　　　　　　(E) $(-1,5)$

4. If $\dfrac{dy}{dx} = y \cos x$ and $y = 3$ when $x = 0$, then $y =$

    (A) $e^{\cos x - 1} + 2$　(B) $x + 3$　　(C) $3e^{\sin x}$　　(D) $\sin x + 3$　(E) $\sin x + 3e^x$

5. If $f'(x) = 2f(x)$ and $f(1) = 1$, then $f(2) =$

    (A) e　　　　(B) 2　　　　(C) e^2　　　　(D) 4　　　　(E) none of these

6. If $\dfrac{dx}{dt} = kt$, and if $x = 2$ when $t = 0$ and $x = 6$ when $t = 1$, then k equals

    (A) $\ln 4$　　(B) 8　　　　(C) e^3　　　　(D) 3　　　　(E) none of these

7. If $\dfrac{dx}{dt} = -10x$ and if $x = 50$ when $t = 0$, then $x =$

    (A) $50 \cos 10t$　　　　(B) $50e^{-10t}$　　　　(C) $50e^{10t}$

    (D) $50 - 10t$　　　　　(E) $50 - 5t^2$

8. If $\dfrac{dy}{dx} = \dfrac{\cos x}{\cos y}$ and $(0,0)$ is a point on the curve, which one of the following points is also on the curve?

(A) $\left(\dfrac{\pi}{3}, \dfrac{2\pi}{3}\right)$ (B) $\left(0, \dfrac{\pi}{2}\right)$ (C) $\left(\dfrac{\pi}{3}, \dfrac{5\pi}{3}\right)$ (D) $\left(\dfrac{\pi}{4}, \dfrac{-\pi}{4}\right)$ (E) $(0,1)$

9. If $\dfrac{dy}{dx} = e^y$ and $y = 0$ when $x = 1$, then $y = -1$ when $x =$

(A) 2 (B) e (C) $2 - e$ (D) $2 + e$ (E) $\ln(2 + e)$

10. The curve that passes through the point $(1, e)$ and whose slope at any point (x, y) is equal to $\dfrac{3y}{x}$ has the equation

(A) $y = x^3$ (B) $y = 3x$ (C) $y = ex^3$

(D) $y = 3x^3$ (E) $y = 3e^x$

11. On the surface of the moon, the acceleration of gravity is -5.28 feet per second per second. If an object is thrown upward from an initial height of 1000 feet with a velocity of 56 feet per second, its velocity 4.5 seconds later is

(A) 67.88   (B) 37.52   (C) 32.24   (D) 25.16   (E) 16.12

12. If the temperature is constant, then the rate of change of barometric pressure p with respect to altitude h is proportional to p. If $p = 30$ in. at sea level and $p = 29$ in. at $h = 1000$ ft, then the pressure at 5000 ft is

(A) 21.47   (B) 25.32   (C) 28.91   (D) 32.11   (E) 35.82

Answers to Worksheet 2
1. C	2. C	3. A	4. C	5. C	6. B
7. B	8. A	9. C	10. C	11. C	12. B

Worksheet 3

1. The temperature inside a refrigerator is maintained at 5 °C. An object at 100 °C is placed in the refrigerator to cool. After 1 minute, its temperature drops to 80 °C. How long would it take for the temperature to drop to 10 °C? Assume Newton's Law of Cooling.

2. At 6 a.m. the temperature of a corpse is 13 °C and 3 hours later it falls to 9 °C. The living body has a temperature of 37 °C. Assuming the temperature of the room in which the body rests is 5 °C, and assuming Newton's Law of Cooling estimate the time of death.

3. When a transistor radio is switched off, the current declines according to the formula $\dfrac{dI}{dt} = kI$ where I is the current, t is time in seconds and k is a constant. If the current drops to 10 % in the first second, how long will it take to fall to 0.1 % of its original value?

4. Water evaporates from a lake at a rate proportional to the volume of water remaining. If 50 % of the water evaporates in 20 days, find the percentage of the original water remaining after 50 days without rain.

5. Let f be a function with $f(1) = 4$ such that for $x > 0$ and $y > 0$, slope is given by $\dfrac{3x^2 + 1}{2y}$.

    a) Find the slope of the graph of f at the point where $x = 1$.

    b) Write an equation for the line tangent to the graph of f at $x = 1$ and use it to approximate $f(1.2)$.

c) Find $f(x)$ by solving the separable differential equations $\dfrac{dy}{dx}=\dfrac{3x^2+1}{2y}$ with

the initial condition $f(1)=4$.

d) Use your solution from part (c) to find $f(1.2)$.

Answers to Worksheet 3

1. 12.46 minutes   2. 12 midnight   3. 3 seconds   4. 17.68 %

5. a) $\dfrac{1}{2}$ b) $y=\dfrac{1}{2}x+\dfrac{7}{2}$, $f(1.2)\approx 4.1$ c) $f(x)=\sqrt{x^3+x+14}$ d) 4.114

Worksheet 4

1. The bacteria in a certain culture increase at a rate proportional to the number present.

   a) If the number triples in 8 hours, how many are there in 12 hours?

   b) In how many hours will the original number quadruple?

2. Let $P(t)$ represent the number of deer in a population at time t years, when $t\geq 0$. The population $P(t)$ is increasing at a rate directly proportional to $300-P(t)$, where the constant of proportionality is k.

   a) If $P(0)=100$, find $P(t)$ in terms of t and k.

   b) If $P(2)=200$, find k.

   c) Find $\lim\limits_{t\to\infty} P(t)$.

3. The thickness, $f(t)$ (inches), of ice forming on a lake satisfies the differential

   equation $f'(t) = \dfrac{3}{f(t)}$, where t is measured in hours.

   a) If $f(0) = 1$, find $f(t)$.

   b) When is the thickness two inches?

4. A roast is put in a 300 °F oven and heats according to the differential equation

$$\frac{dT}{dt} = k(300 - T)$$

   where k is a positive constant and $T(t)$ is the temperature of the roast after t

   minutes.

   a) If the roast is at 50 °F when put in the oven, i.e. $T(0) = 50$, find $T(t)$ in

      terms of k and t.

   b) If $T(30) = 200$ °F, find k.

5. A certain population increases at a rate proportional to the square root of the

   population. If the population goes from 2500 to 3600 in five years, what is it

   at the end of t years? Assume that at $t = 0$ years the population is 2500.

6. The rate of change of volume V of a melting snowball is proportional to the

   surface area of the snowball, that is, $\dfrac{dV}{dt} = -kS$, where k is a positive constant.

   If the radius of the ball at $t = 0$ is $r = 2$ and at $t = 10$ is $r = 0.5$, show that

   $r = -\dfrac{3}{20}t + 2$.

7. Let $F(t)$ be the temperature, in degrees Fahrenheit, of a cup of tea at time t minutes, $t \geq 0$. Room temperature is $70\,°$ and the initial temperature of the tea is $180\,°$. The tea's temperature at time t is described by the differential equation $\dfrac{dF}{dt} = -0.1(F-70)$, with the initial condition $F(0)=180$.

a) Find an expression for F in terms of t, where t is measured in minutes.

b) How hot is the tea after 10 minutes?

c) If the tea is safe to drink when its temperature is less than $120\,°$, at what time is the tea safe to drink?

Answers to Worksheet 4

1. a) 5.196 times as many   b) 10 hours 6 minutes

2. a) $P(t)=300-200e^{-kt}$ b) $k=\dfrac{1}{2}\ln 2$ c) 300

3. a) $\sqrt{6t+1}$ b) after $\dfrac{1}{2}$ hour

4. a) $H(t)=300-250e^{-kt}$ b) 0.03054

5. $(2t+50)^2$

7. a) $F = 70+110e^{\frac{-t}{10}}$

b) $110\,°$ (approx.)

c) 7.88 minutes

Worksheet 5

1. The half-life of a substance is 3 minutes. Of an initial amount of 100 grams, how much will remain after 20 minutes?

2. Polonium 210 decays into lead with a half-life of 138 days. How long will it take for 90 % of the radioactivity in a sample of Polonium 210 to dissipate?

3. A bowl of soup initially at 80 °C cools to 40 °C in 15 minutes when the room temperature is 20 °C. Roger refuses to drink his soup if it cools to a temperature below 50 °C. How long does he have before he needs to come to the table for his soup?

4. An indoor thermometer reading 20 °C is put outdoors. In 10 minutes it reads 25 °C and in another 10 minutes it reads 27 °C. Calculate the outdoor temperature.

5. What annual rate of interest compounded annually is equivalent to an annual rate of 6 % compounded continuously?

6. A cook monitors the temperature of a roast that is in a 200 °C oven. At 2 p.m., the temperature of the roast is 80 °C and at 3 p.m. the temperature of the roast is at 156 °C. If the temperature of the roast was initially 20 °C at what time was the roast put in the oven?

7. The motion of a stone falling from rest from the top of a high building is governed by the differential equation

$$\text{acceleration} = \frac{dv}{dt} = 32 + \frac{1}{2}v$$

where v and t represent velocity and time respectively as usual. Find the velocity after 4 seconds. Find also the distance the stone falls from its initial position after 2 seconds.

8. When a person dies the temperature of their body will decrease from 37 °C to the temperature of the surroundings. The situation is modeled by Newton's Law of Cooling. A person was found murdered. Police arrived at the scene of the murder at 10:56 p.m. The temperature of the body at that time was 31 °C and one hour later it was 30 °C. The temperature of the room in which the murder was committed is 22 °C. At what time (to the nearest minute) was the murder committed?

9. The population of Russia in 1969 was 209 million. It is estimated that the population P is increasing exponentially at a rate of 1 % per year i.e.

$$\frac{dP}{dt} = .01P.$$

   a) Estimate the population of Russia in 2009.

   b) After what period of time will the population be double that of 1969?

10. Research has provided data which substantiate the model that the risk (r %) of having an automobile accident is related to the blood alcohol level (b %) by

$$\frac{dr}{db} = kr$$ where k is a constant. The risk is 1% if the blood alcohol level is 0 % and the risk is 20 % if the blood alcohol level is 14%. At what blood alcohol level will the risk of having an accident be 80%?

11. When a coil of steel is removed from a furnace its temperature is 684 °C. Four minutes later its temperature is 246 °C. How long will it take for the coil of steel to cool to 100 °C? The surrounding temperature is 27 °C. Assume Newton's Law of Cooling.

12. The velocity of a skydiver satisfies the differential equation

$$\frac{dv}{dt} = \left(g - \frac{1}{4}v \right)$$

where g is the acceleration due to gravity.

a) Find the velocity as a function of time. Assume initial velocity is 0.

b) Show that the limiting velocity of the skydiver is $4g$.

c) How long will it take for the skydiver to reach 90% of his limiting velocity (often called terminal velocity?)

13. Let $v(t)$ be the velocity, in feet per second, of a skydiver at time t seconds, $t \geq 0$. After her parachute opens, her velocity satisfies the differential equation $\frac{dv}{dt} = -2v - 32$, with initial condition $v(0) = -50$.

a) Find an expression for v in terms of t, where t is measured in seconds.

b) Terminal velocity is defined as $\lim_{t \to \infty} v(t)$. Find the terminal velocity of the skydiver to the nearest foot per second.

c) It is safe to land when her speed is 20 feet per second. At what time does she reach this speed?

Answers to Worksheet 5

1. 0.984 grm

2. 458.4 days

3. 9.46 minutes

4. $28\frac{1}{3}$ °C

5. 6.184 %

6. 1.36 p.m.

7. 408.9, 91.94

8. 6:36 p.m.

9. a) 311,791,000

 b) 69 years

10. 20.5 %

11. 8 minutes

12. a) $v = 4g\left(1 - e^{\frac{-t}{4}}\right)$

 c) 9.2 seconds (approx.)

13. a) $|v + 16| = 34e^{-2t}$

 b) −16 ft/sec

 c) 1.07 seconds

Slope Fields

A slope field is a relatively recent method for helping to solve differential equations.

It involves drawing, either by hand, by graphing calculator, or by computer, a picture

of small segments representing the slopes at many points as calculated by substituting

into the differential equation given.

For example, if we were given $\dfrac{dy}{dx} = -\dfrac{x}{y}$ (and $y = 2$ when $x = 0$) then the slope field

would look like

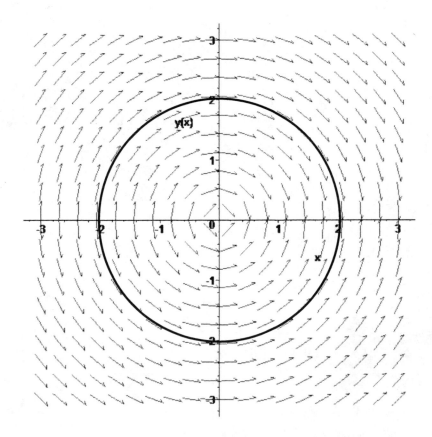

Note, for example, that at the point (1,1) the slope is $\dfrac{-1}{1} = -1$ and at (-1,1) the slope

is $\dfrac{--1}{1} = \dfrac{1}{1} = 1$.

It must be remembered that the slope field is a representation of the <u>original</u> equation and as such is a representation of the <u>solution</u> to the differential equation. For our example it certainly seems that the solution to the differential equation is a circle. We can check this by solving the differential equation analytically:

$$\frac{dy}{dx} = -\frac{x}{y}$$

$$\therefore \qquad y\frac{dy}{dx} = -x$$

$$\therefore \qquad \frac{1}{2}y^2 = -\frac{x^2}{2} + c$$

i.e. $\quad y^2 + x^2 = 2c$.

When $x = 0$, $y = 2$, $\therefore c = 2$.

i.e. $\quad y^2 + x^2 = 4$ which is a circle centre (0,0) as shown on the slope field.

Of course greater value obtains to using slope fields when we are trying to solve more complex differential equations.

For example, consider the differential equation $\dfrac{dy}{dx} = x + y$.

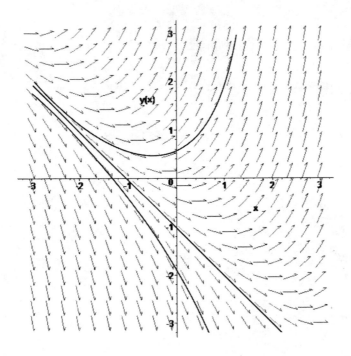

Shown are the representations of the solutions passing through

1) $(0, \dfrac{1}{2})$ 2) $(0,-1)$ and   3) $(0,-2)$

Solving $\dfrac{dy}{dx} = x + y$ analytically is beyond the scope of this book but suffice it to note

that the solution is $y = -x - 1 + ce^{x}$ for some constant c which is determined by being

told a point through which the graph passes.

Note for example that if the graph passes through $(0,1)$ then the solution is

$y = -x - 1 + 2e^{x}$ which is modelled by the top-most graph outlined on the picture.

Similarly if the graph passes through $(0,-1)$ then the solution is $y = -x - 1$ which is

clearly indicated by the picture.

Furthermore if the graph passes through (0,-2) then the solution is $y=-x-1-e^x$

which is modelled by the bottom-most graph outlined in the picture.

Remember that the solution to differential equations is modelled by starting at the

point given and tracing on the slope field the direction suggested by the line segment

slopes. For example below is shown the slope field for the solution to:

$$\frac{dy}{dx}=-y-x^2$$

Drawn in thick type are the solutions (top to bottom) passing through

a) (0,2)　　　b) (0,0) and　c) (0,-2)

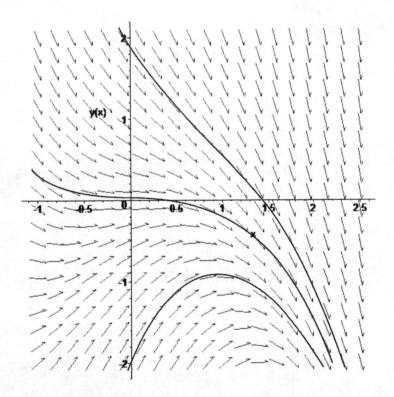

Again, analytically solving for the solutions to the differential equations is beyond the

scope of this book but for your interest the solutions are

a) $y=-x^2+2x-2+4e^{-x}$　　　b) $y=-x^2+2x-2+2e^{-x}$　　　c) $y=-x^2+2x-2$

which are represented on the picture.

Worksheet 7

1. Match the slope fields shown below with their differential equations.

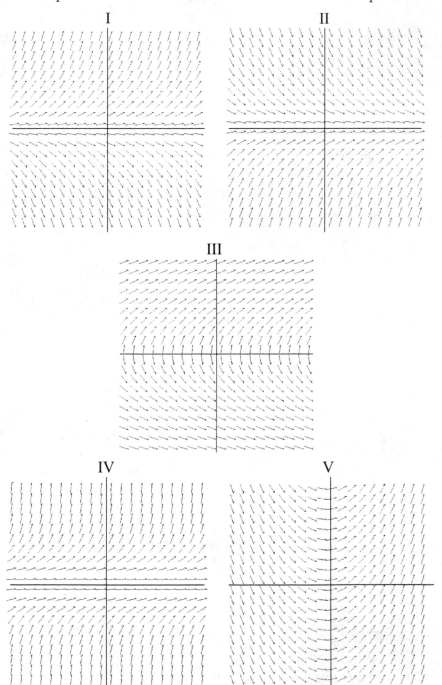

I

II

III

IV

V

(A) $\dfrac{dy}{dx} = -y$ (B) $\dfrac{dy}{dx} = y$ (C) $\dfrac{dy}{dx} = x$ (D) $\dfrac{dy}{dx} = \dfrac{1}{y}$ (E) $\dfrac{dy}{dx} = y^2$

2. Match the slope fields shown below with their differential equations.

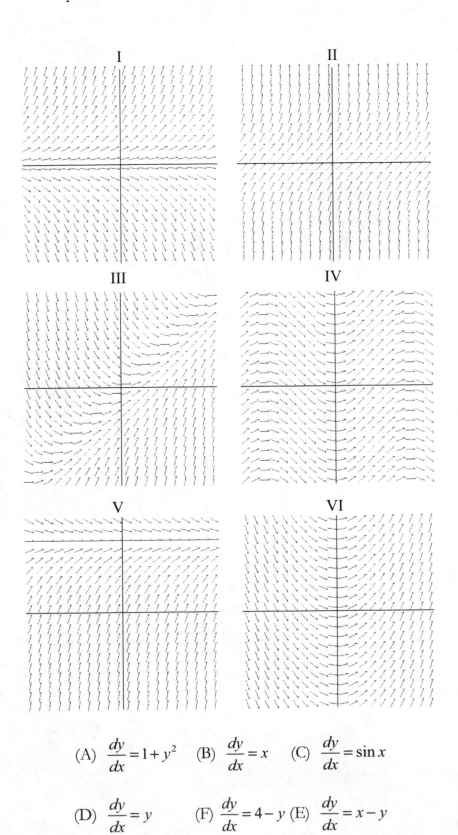

I

II

III

IV

V

VI

(A) $\dfrac{dy}{dx} = 1 + y^2$ (B) $\dfrac{dy}{dx} = x$ (C) $\dfrac{dy}{dx} = \sin x$

(D) $\dfrac{dy}{dx} = y$ (F) $\dfrac{dy}{dx} = 4 - y$ (E) $\dfrac{dy}{dx} = x - y$

3. The slope field for the differential equation $\dfrac{dy}{dx} = y - x^2$ together with two

special solutions is shown below.

a) Verify that $y = x^2 + 2x + 2$ and $y = x^2 + 2x + 2 - 2e^x$ are the two solutions

shown and indicate which solution goes with which curve.

b) Estimate the y-intercept of the solution which passes through (-1,0).

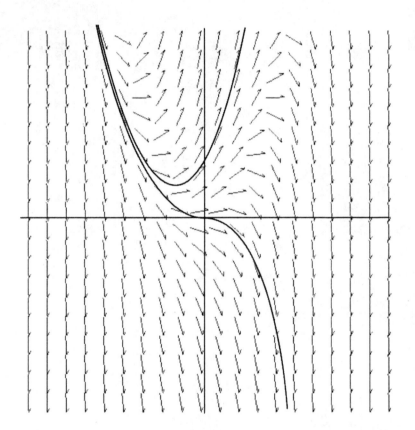

4. Which of the slope fields shown below represents $\dfrac{dy}{dx} = \dfrac{x+y}{x-y}$

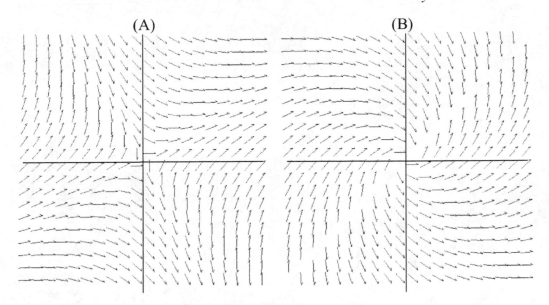

(A)   (B)

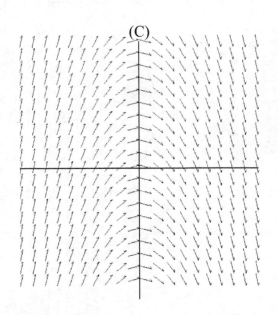

(C)

5. Match the slope fields shown below with the following differential equations.

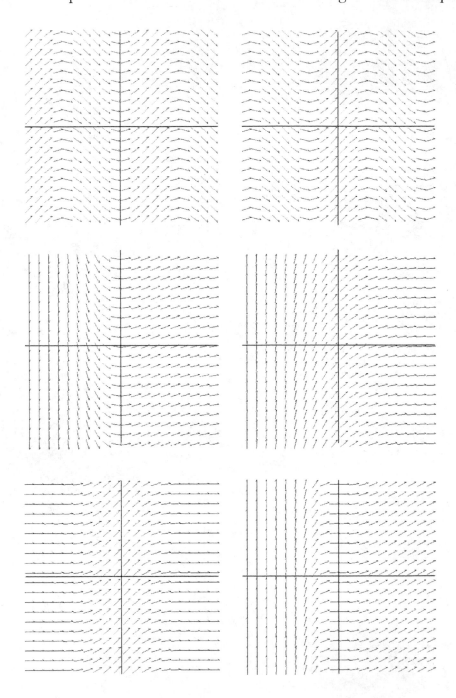

(A) $\dfrac{dy}{dx} = xe^{-x}$ (B) $\dfrac{dy}{dx} = \sin x$ (C) $\dfrac{dy}{dx} = \cos x$

(D) $\dfrac{dy}{dx} = x^2 e^{-x}$ (E) $\dfrac{dy}{dx} = e^{-x^2}$ (F) $\dfrac{dy}{dx} = e^{-x}$

6. Shown below is a slope field for the differential equation $\dfrac{dy}{dx} = \dfrac{(y+1)(2-y)}{2}$.

    a) Plot the points   i) $(0,0)$

                             ii) $(0,1)$

                             iii) $(1,0)$

   on the slope field.

    b) Plot solution curves through the points in a).

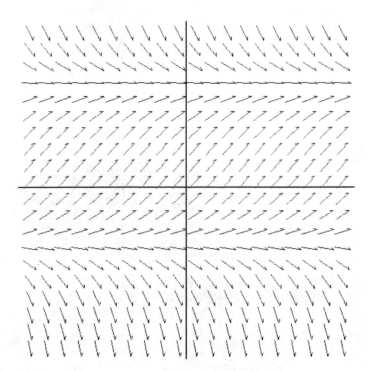

7. Consider the differential equation $\dfrac{dy}{dx} = 2y - 4x$.

   a) The slope field for the given differential equation is provided. Sketch the solution curve that passes through the point (0,1) and sketch the solution curve that passes through the point (0,-1).

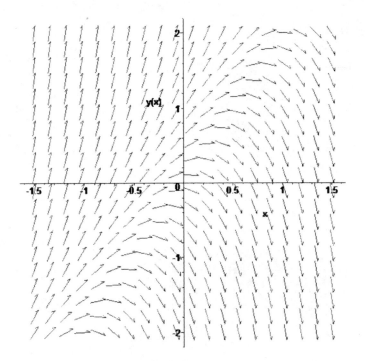

   b) Let f be the function that satisfies the differential equation with initial condition $f(0)=1$. Use Euler's method, starting at $x = 0$ with a step size of 0.1, to approximate $f(0.2)$.

   c) Estimate the value of b for which $y = 2x + b$ is a solution to the given differential equation.

   d) Let g be the function that satisfies the given differential equation with the initial condition $g(0)=0$. Does the graph of g have a local extremum at the point (0,0)? If so, is the point a local maximum or a local minimum?

8. The figure below shows the slope field for a differential equation $\dfrac{dy}{dx} = f(x)$.

Let $g(x) = \int f(x)\,dx + C$ be the family of functions which are solutions of the differential equation.

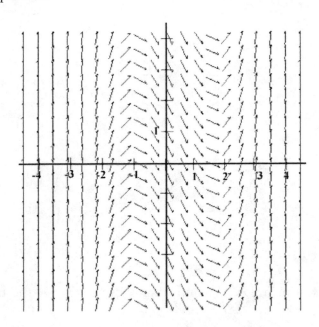

a) Determine to the nearest integer the value of x for which all of the members of the family of $g(x)$ will have a relative minimum value. Justify your answer.

b) Determine to the nearest integer the value of x for which all of the members of the family of $g(x)$ will have a relative maximum value. Justify your answer.

c) Sketch the member of the family of $g(x)$ for which $g(0) = -2$. Justify your answer.

d) For the function sketched in part c), determine the solution(s) of $g(x) = 0$ to the nearest integer.

Answers to Worksheet 7

1. I – B, II – A, III – D, IV – E, V – C

2. I – D, II – A, III – E, IV – C, V – F, VI – B

3. a) Upper graph is $y = x^2 + 2x + 2$

Lower graph is $y = x^2 + 2x + 2 - 2e^x$

b) (0,-1)

4. B

5. I – B, II – C, III – A, IV – F, V – E, VI – D

7. b) $f(0.2) \approx 1.4$ c) $b = 1$ d) Yes, local maximum.

8. a) $x = 2$. To the left of $x = 2$ the slope is negative, to the right it is positive. The functions are decreasing, then increasing. By the First Derivative Test this indicates a relative minimum near $x = 2$.

b) $x = -1$. To the left of $x = -1$ the slope is positive, to the right it is negative. The functions are increasing, then decreasing. By the First Derivative Test this indicates a relative maximum near $x = -1$.

c) Start at the point (0,-2) and sketch in both directions, following the directions indicated by the segments which make up the slope field.

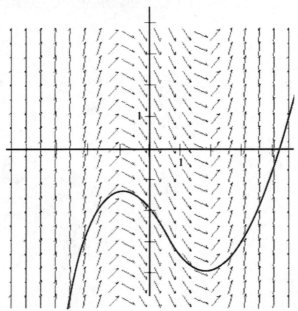

d) The sketch intersects the x-axis near $x = 4$. This is (approximately) the solution of the equation graphed in part a).

Note that since the derivative is a function for x <u>only</u>, the slope will be the same for each member of the family, at each value of x. This is why each segment of the slope field appears parallel to those segments above and below it. This is also why maximums and minimums occur at the same x value for all members of the family.

CHAPTER 16

Volumes of Solids of Revolution

In an earlier chapter we saw how integration can be used to evaluate areas of certain

regions. Similarly integration can be used to find volumes of solids formed by

rotating a region about a given straight line. The method used is analogous to the

technique of using thin strips to find areas.

For example consider the region R bounded by $x + 2y = 4$, the x-axis and the

y-axis:

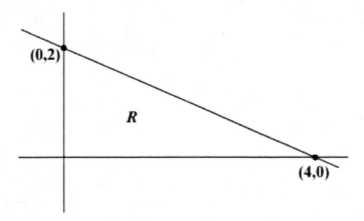

When R is rotated $360°$ about the y-axis a cone is formed as shown below:

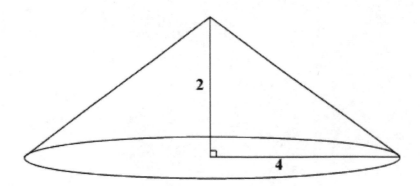

To find the volume of the cone we could simply use the formula $V = \frac{1}{3}\pi r^2 h$

i.e. $\frac{32}{3}\pi$ but instead we will use this as an introductory example of the "strip"

method for finding volumes.

Consider a thin horizontal strip as shown below whose length is x and whose width

is dy.

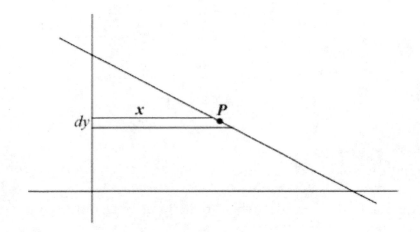

Remember that x refers to the x co-ordinate of a point (P) on the graph and

therefore does represent the length of the mini strip.

When the thin strip is rotated $360\,°$ about the y-axis it forms a thin disc whose

volume is $\pi x^2 dy$.

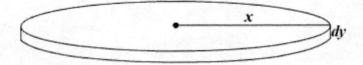

Note that the cone is composed of an infinite number of thin horizontal strips whose

sum will yield the cone. It follows that the volume of the cone will be $\int_0^2 \pi x^2 dy$.

Hence the volume of the cone $\quad = \int_0^2 \pi x^2 dy$

$$= \pi \int_0^2 (4-2y)^2 \, dy$$

$$= \pi \left[-\frac{1}{6}(4-2y)^3 \right]_0^2$$

$$= \pi \left[0 + \frac{64}{6} \right] = \frac{32\pi}{3} \text{ as previously shown.}$$

When evaluating the volume of a solid of revolution it is important to concentrate

upon what happens to the thin strip when rotated rather than the whole solid itself.

Example

DISC METHOD

Question: The region bounded by $y = x^2$, $x = 2$, $x = 3$, and the x axis, is rotated

360 ° about the x-axis. Find the volume of the figure so formed.

Answer:

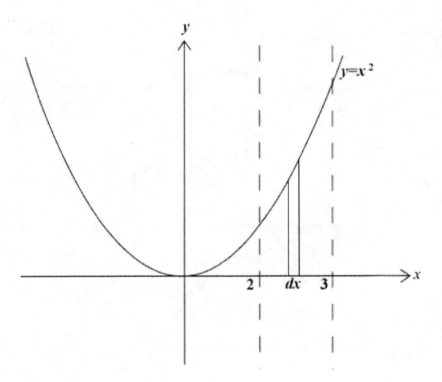

Consider a thin vertical strip width dx, height y. When it is rotated

about the x-axis it forms a disc whose volume is $\pi y^2 dx$.

The volume of the solid $= \int_{2}^{3} \pi y^2 dx$

$$= \pi \int_{2}^{3} x^4 dx = \pi \left[\frac{1}{5} x^5 \right]_{2}^{3}$$

$$= \frac{211\pi}{5} = 132.575 \text{ (approx.)}$$

Example

WASHER METHOD

Question: The region bounded by $y = x^3$ and $y = 2x^2$ is rotated 360° about the

x -axis. Find the volume of the solid so formed.

Answer:

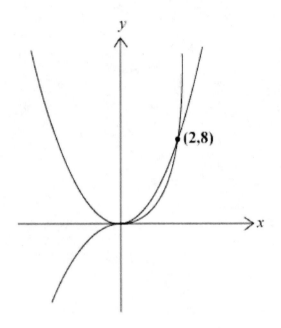

$y = 2x^2$ intersects $y = x^3$ at (2,8) and (0,0). We wish to find the

volume of the "crescent" shaped region rotated around the x -axis.

Consider a thin vertical strip width dx.

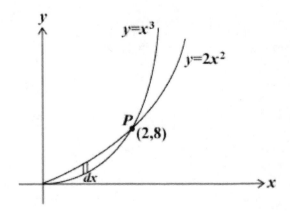

When it is rotated around the x-axis it becomes a "washer"

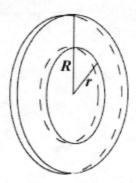

whose volume is $\pi R^2 dx - \pi r^2 dx$ where R, r

refer to the radii of the large and small circle respectively.

Note that $R = 2x^2$ and $r = x^3$.

Volume of the washer
$$= \pi \left(2x^2\right)^2 dx - \pi \left(x^3\right)^2 dx$$

$$= \pi \left(4x^4 - x^6\right) dx$$

$\therefore$ Volume of solid of revolution
$$= \pi \int_0^2 4x^4 - x^6 dx$$

$$= \pi \left[\frac{4}{5}x^5 - \frac{x^7}{7}\right]_0^2 = \frac{256\pi}{35} = 22.979$$

Example

In the previous example, if the region had been rotated about the y-axis then

its volume could have been found as follows.

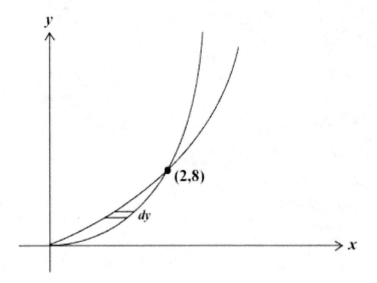

Consider a thin horizontal strip which, when rotated about the y-axis, would

again form a "washer" whose volume is $\pi P^2 dy - \pi p^2 dy$ where P, p are the

large and small radii respectively.

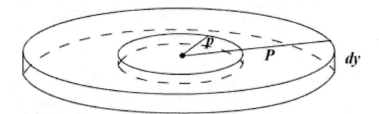

P is the x co-ordinate of a point on $y = x^3$ and p is the x co-ordinate of a

point on $y = 2x^2$. i.e. $P = y^{\frac{1}{3}}$ and $p = \dfrac{y^{\frac{1}{2}}}{\sqrt{2}}$.

The volume of the washer $\quad = \pi y^{\frac{2}{3}} dy - \pi \frac{y}{2} dy$

The volume of the solid of revolution $\quad = \pi \int_{0}^{8} y^{\frac{2}{3}} - \frac{y}{2} dy = \pi \left[\frac{3}{5} y^{\frac{5}{3}} - \frac{y^2}{4} \right]_{0}^{8}$

$$= \frac{16\pi}{5} = 10.053 .$$

Frequently it is possible to consider either horizontal or vertical strips but in certain instances it is either impractical or impossible to use one of them.

Example

Question: Consider the region R bounded by $y = -x^2 + 3x$ and $y = x$. Find the volume of the solid formed when this region R is rotated about the x-axis.

Answer:

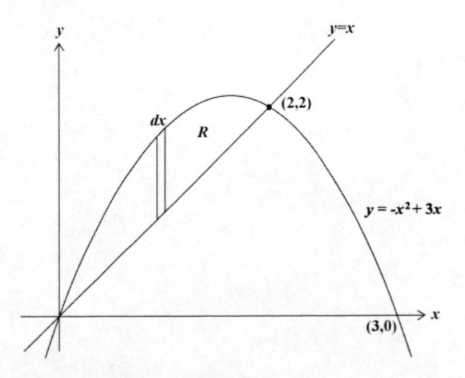

Clearly it is impractical to use horizontal strips for values of y greater than 2 and hence a vertical strip is used. The thin vertical strip, when rotated about the x-axis produces a washer whose volume is

$$\pi \left(-x^2 + 3x\right)^2 dx - \pi \left(x\right)^2 dx.$$

The volume of the solid of revolution:

$$= \pi \int_0^2 \left(-x^2 + 3x\right)^2 - x^2 dx$$

$$= \pi \int_0^2 x^4 - 6x^3 + 8x^2 dx$$

$$= \pi \left[\frac{x^5}{5} - \frac{3x^4}{2} + \frac{8x^3}{3}\right]_0^2$$

$$= \frac{56}{15}\pi = 11.729 \text{ (approx.)}$$

Example

SHELL METHOD

Question: Consider the solid formed when the region R in the last example is rotated about the y-axis. Find the volume of the solid so formed.

Answer: As explained it is impractical to use a horizontal strip and hence the thin vertical strip, when rotated 360 ° about the y-axis produces a shape known as a shell whose dimensions are shown in the figure below. The word shell is used in its military, rather than marine, sense.

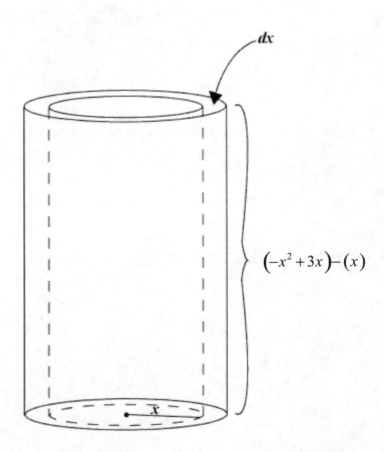

$$\left(-x^2+3x\right)-\left(x\right)$$

To evaluate the volume of the shell so formed, imagine cutting open the shell to produce a lamina as shown below

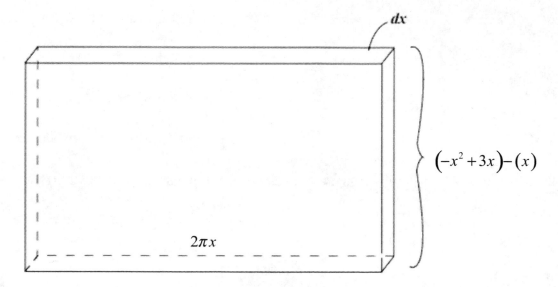

whose dimensions are as shown.

The volume of the lamina is $(2\pi x)(-x^2 + 2x)dx$.

The volume of the solid of revolution formed is

$$\int_0^2 2\pi x\left(-x^2 + 2x\right)dx$$

$$= 2\pi\int_0^2 \left(-x^3 + 2x^2\right)dx$$

$$= 2\pi\left[-\frac{x^4}{4} + \frac{2}{3}x^3\right]_0^2$$

$$= \frac{8\pi}{3}$$

$$= 8.378$$

Example

Question: The region bounded by the x-axis and $y = 2x - x^2$ is rotated around

the vertical line $x = 3$. Find the volume of the solid so formed.

Answer:

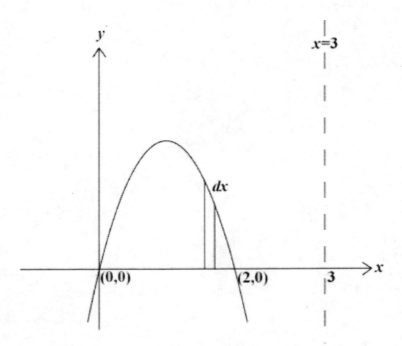

When rotated the vertical thin strip becomes a solid as shown below.

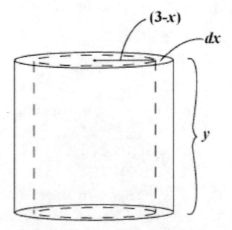

The volume of this thin shell considered as a lamina when cut open is

$$2\pi(3-x)ydx$$

$$=2\pi(3-x)(2x-x^2)dx$$

Total volume of solid of revolution

$$=\int_0^2 2\pi(3-x)(2x-x^2)dx$$

$$=2\pi\int_0^2(6x-5x^2+x^3)dx$$

$$=2\pi\left[3x^2-\frac{5}{3}x^3+\frac{x^4}{4}\right]_0^2$$

$$=2\pi\left[\frac{8}{3}\right]=\frac{16\pi}{3}.$$

Worksheet 1

1. Find the volume of the solid of revolution formed by rotating the

   ellipse $x^2+\dfrac{y^2}{4}=1$ a) about the x-axis.   b) about the y-axis.

2. Find the volume of the solid generated by rotating about the x-axis, the area formed by $y=x^2$, the x-axis, and $x=1$.

3. Find the volume of the solid generated by rotating about the line $x=2$, the area bounded by $y=x^2$, the x-axis, and $x=2$.

4. O is the origin, P is (2,0), N is (2,2), and M is (0,2).

   a) Find the volume of the solid of revolution formed by rotating triangle

ONP about the x-axis.

b) Find the volume of the solid of revolution formed by rotating triangle

ONM about the x-axis.

c) Since clearly area of ΔONM = area of ΔONP, why are the volumes

generated different?

5. Find the volume of the solid generated by rotating the region bounded by

$y = x^2$ and $y = 4x - x^2$ a) about the x-axis and b) about the line $x = 5$.

6. The graph of $y^2 = x^2(12 - x)$ is drawn below.

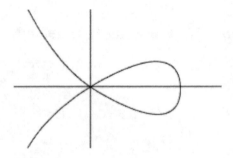

Find the volume of the solid generated by rotating the loop around the x-axis.

7. The area bounded by the curve $y = x - \dfrac{1}{x}$, the x-axis, $x = 2$, and $x = 3$ is

rotated about the x-axis. Find the volume of the solid figure so formed.

Answers to Worksheet 1

1. a) $\dfrac{16\pi}{3}$ b) $\dfrac{8\pi}{3}$ 2. $\dfrac{\pi}{5}$ 3. $\dfrac{8\pi}{3}$ 4. a) $\dfrac{8\pi}{3}$ b) $\dfrac{16\pi}{3}$

5. a) $\dfrac{32\pi}{3}$ b) $\dfrac{64\pi}{3}$ 6. 1728π

7. $\dfrac{9\pi}{2}$

Worksheet 2 – Calculators Permitted

1. The region R is bounded above by the graph of $xy = 1$, on the left by $x = 1$, on the right by $x = 2$, and below by $y = 0$. The volume of the solid of revolution formed when R is revolved about the line $x = 3$ is nearest in value to:

(A) 6.8   (B) 7.0   (C) 7.2   (D) 7.4   (E) 7.6

2. The region R in the first quadrant is bounded by $f(x) = 4 - x$ and $g(x) = \dfrac{3}{x}$. If R is revolved about the y-axis, the volume of the solid formed is nearest in value to:

(A) 8.16   (B) 8.26   (C) 8.36   (D) 8.38   (E) 8.42

3. Let S be the region in the first quadrant bounded by the x-axis, the line $x = 1$ and the graph of $y = \sqrt{\operatorname{Arc\,sin} x}$. What is the volume of the solid generated by rotating S about the y-axis?

(A) 2.67   (B) 2.70   (C) 2.73   (D) 2.76   (E) 2.79

4. Let R be the region in the first quadrant bounded by the x-axis and the curve $y = 2x - x^2$. The volume produced when R is revolved about the x-axis is:

(A) $\dfrac{16\pi}{15}$ (B) $\dfrac{8\pi}{3}$ (C) $\dfrac{4\pi}{3}$ (D) 16π (E) 8π

5. Let R be the region in the first quadrant bounded above by the graph of

   $f(x) = 2\operatorname{Arc}\tan x$ and below by the graph of $y = x$. What is the volume of

   the solid generated when R is rotated about the x-axis?

   (A) 1.21   (B) 2.68   (C) 4.17   (D) 6.66   (E) 7.15

6. Let R be the region enclosed by the graphs of $y = e^{\frac{x}{3}}$, $y = \dfrac{x}{3} + 1$, and the line

   $x = 4$. The volume of the solid generated when R is revolved about the

   y-axis is nearest to

   (A) 33.09   (B) 33.11   (C) 33.13   (D) 33.15   (E) 33.17

7. Let R be the region in the first quadrant enclosed by the graphs of $y = x^{\sin x}$

   and the lines $x = 1$ and $x = 3$. What is the volume of the solid generated when

   R is rotated about the y-axis?

   (A) 15.9   (B) 18.7   (C) 40.1   (D) 50.6   (E) 64.9

8. Which definite integral represents the volume of a sphere with radius 2?

   (A) $\pi \displaystyle\int_{-2}^{2} (x^2 - 4)\,dx$ (B) $\pi \displaystyle\int_{-2}^{2} (x^2 + 4)\,dx$ (C) $2\pi \displaystyle\int_{0}^{2} (4 - x^2)\,dx$

   (D) $2\pi \displaystyle\int_{-2}^{2} (4 - x^2)\,dx$ (E) $\pi \displaystyle\int_{0}^{2} (4 - x^2)\,dx$

Answers to Worksheet 2

   1. A   2. D   3. B   4. A   5. E   6. B   7. C   8. C

Volumes of Solids with Known Cross-Section

It is important to remember that integration is the process of adding together an infinite number of small "things". For example suppose we had a closed vessel whose cross-section area, parallel to the base, was $(36-x^2)$ square metres where x represented the distance in metres of the cross-section region from the base.

The base area, when $x=0$, would be 36 square metres and the height of the vessel would be 6 metres because, when $x=6$ the cross-section area equals zero. The vessel would resemble a "beehive" as shown.

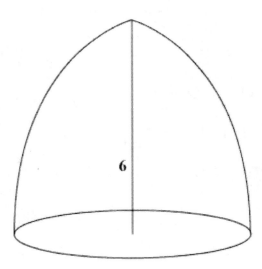

If we wished to find the volume of the "beehive" then we would simply evaluate

$\int_0^6 (36-x^2)dx$ since $(36-x^2)dx$ represents the volume of a thin horizontal strip and

the integration process adds them all up.

$$\text{Volume} = \left[36x - \frac{x^3}{3}\right]_0^6 = 144 \text{ cubic metres.}$$

Similarly consider the following question taken from an Advanced Placement

Calculus AB examination.

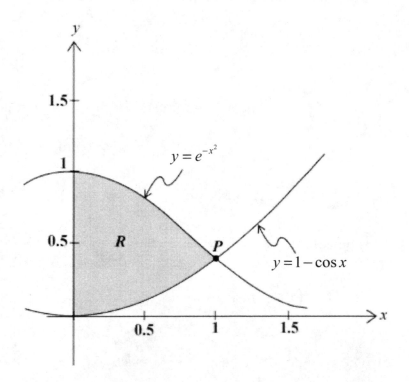

Let R be the region in the first quadrant enclosed by the graphs of $y = e^{-x^2}$,

$y = 1 - \cos x$, and the y-axis, as shown in the figure above.

a) Find the area of the region R.

b) Find the volume of the solid generated when the region R is revolved about the

   x-axis.

c) The region R is the base of a solid. For this solid, each cross section

   perpendicular to the x-axis is a square. Find the volume of this solid.

Answer:

a) We need to find first the point of intersection P of $y = e^{-x^2}$ and

$y = 1 - \cos x$. By calculator, P is $(0.94194408, 0.41178305)$.

By considering a thin vertical strip it is clear that the area of

region $R = \displaystyle\int_0^{0.94194408} e^{-x^2} - (1 - \cos x)\,dx = 0.591$ (approx.)

b) Similarly by considering a thin vertical strip, when rotated around

the x-axis, it becomes a washer whose volume is

$$\pi\left(e^{-x^2}\right)^2 dx - \pi\left(1 - \cos x\right)^2 dx.$$

The volume of the solid of revolution so formed is

$$\pi \int_0^{0.94194408} \left(e^{-2x^2} - 1 + 2\cos x - \cos^2 x\right)dx = 1.7466 \text{ (approx.)}$$

c) The volume of each square cross-section is $\left[e^{-x^2} - (1 - \cos x)\right]^2 dx$.

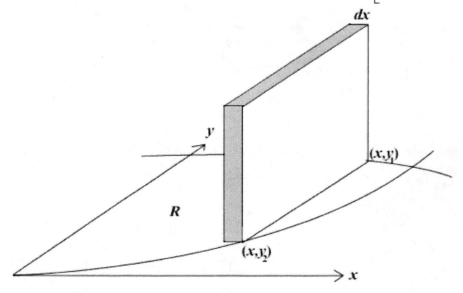

and hence the total volume of the solid is

$$\int_0^{0.94194408} \left(e^{-x^2} - 1 + \cos x\right)^2 dx = 0.4611 \text{ (approx.)}$$

Example

Question: Shown below is the base of a solid represented by $x^2 + y^2 \leq 1$. The solid is formed by having cross-sections in planes perpendicular to the y-axis between $y = -1$ and $y = 1$ and are isosceles right-angled triangles with one side in the base. Find the volume of the solid.

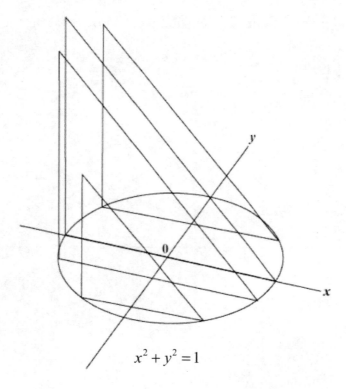

$$x^2 + y^2 = 1$$

Answer: Each cross section has width dy and area $\dfrac{(2x)^2}{2}$.

$\therefore$ each cross section has volume $2x^2 dy$.

The total volume is hence $\displaystyle\int_{-1}^{+1} 2x^2\, dy = 2\int_{-1}^{+1} \left(1 - y^2\right) dy$

$$= 2\left[y - \frac{y^3}{3} \right]_{-1}^{+1} = \frac{8}{3}$$

An interesting theorem concerning rotations of regions is Pappus' Theorem which states that the volume of a solid formed by rotating a region R about a line, not intersecting the region, equals the area of the region R times the distance travelled by the centre of gravity of the region in the rotation.

For example, consider the volume of the solid formed by rotating the circle $x^2 + y^2 = 25$ about the line $x = 8$.

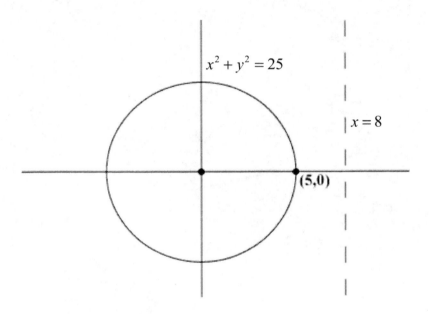

The volume of the torus (doughnut) so formed

$\quad$ = (Area of circle) times (circumference of circle formed when (0,0) rotates

$\qquad$ 360 ° around $x = 8$)

$\quad$ = $\pi 5^2$ times $2\pi 8$

$\quad$ = $400\pi^2$

$\quad$ = 3947.8 (approx.)

Worksheet 3

1. The area bounded by the x-axis, the y-axis, $x = 3$ and $y = \sqrt{x+1}$ is rotated about the y-axis. Find the volume of the solid so formed.

2. A hemispherical bowl of radius 5 cm contains water to a depth of 3 cm. Find the volume of the water.

3. The area in the first quadrant bounded by $y = x$, $y = 2 - x^2$, and the x-axis is rotated about the y-axis. Find the volume of the solid generated.

4. Prove that the volume of a sphere is $\dfrac{4}{3}\pi r^3$ considering the rotation of the circle $x^2 + y^2 = r^2$ about the x-axis.

5. A container is such that its cross-section area is $\left(4 - x^2\right)$ cm² where x is the distance in cm from the base.
   a) What is the height of the container?
   b) What is its volume?

6. The area of a cross-section of a vase at a distance x cm below the top is $\left(4 - x^2\right)$ cm². Find the depth of the water when the vase is half full.

7. Find the volume of the solid generated by rotating the area bounded by $y = x^2$

   and $y = x$

   a) about the x-axis.

   b) about the y-axis.

8. Find the volume generated when the region enclosed by the lines $y = 1$, $x = 3$,

   and $y = \sqrt{x}$ between $(1,1)$ and $(3, \sqrt{3})$ is rotated about the x-axis.

9. A closed vessel tapers to point A and B at its ends and is such that its cross-

   section area cut by a plane perpendicular to AB, x cm from A, is

   $x^2(4-x)$ cm². Find the volume of the vessel.

10. The area bounded by the x-axis, $x = 2$, and $y = \sqrt{x}$ is rotated about the

    y-axis. Find the volume of the solid generated.

11. Find the volume formed by rotating the area bounded $y = 4x^2$, the y-axis

    and $y = 16$,

    a) about the y-axis.

    b) about the line $y = 16$.

12. A solid is 12 inches high. The cross-section of the solid at height x above its base has area $3x$ square inches. Find the volume of the solid.

13. A solid extends from $x=1$ to $x=3$. The cross-section of the solid in the plane perpendicular to the x-axis is a square of side x. Find the volume of the solid.

14. A solid is 6 ft high. Its horizontal cross-section at height x ft above the base is a rectangle with length $(2+x)$ ft and width $(8-x)$ ft. Find the volume of the solid.

15. A solid extends along the x axis from $x=1$ to $x=4$. Its cross-section at any point x is an equilateral triangle with edge $\sqrt{x}$. Find the volume of the solid.

Answers to Worksheet 3

1. 48.59   2. 36π 3. $\dfrac{7\pi}{6}$ 4. ----   5. a) 2   b) $\dfrac{16}{3}$

6. 1.3054   7. a) $\dfrac{2\pi}{15}$ b) $\dfrac{\pi}{6}$ 8. 2π 9. $\dfrac{64}{3}$

10. $\dfrac{16\sqrt{2}}{5}\pi$ 11. a) 32π b) $\dfrac{4096\pi}{15}$ 12. 216   13. $\dfrac{26}{3}$

14. 132   15. $\dfrac{15\sqrt{3}}{8}$

Worksheet 4

1. The finite area in the first quadrant bounded by the curves $y = \dfrac{4}{x}$, $y = \dfrac{4}{x^2}$ and the line $x = 2$ is rotated once about the x-axis. Find the volume of the solid formed.

2. Find the area enclosed by $y = \sqrt{x}$, $x^2 + y^2 = 20$ and the x-axis.

3. Find the volume of the solid formed by rotating the circle $(x-2)^2 + y^2 = 1$ about the y-axis. (Pappus' Theorem required)

4. The area bounded by the x-axis, $x = 2$ and $y = \sqrt{x}$ is rotated about the line $x = 3$. Find the volume of the solid generated.

5. Find the volume of the solid generated by rotating about the y-axis, the region, in the first quadrant, bounded by $y = x^3$ and $y = x$.

6. Find the volume of the solid formed by rotating the area bounded by $y = x^2$, the y-axis and $y = 1$ about the line $y = 2$.

7. Find the volume of the solid formed by rotating $\dfrac{x^2}{25} + \dfrac{y^2}{16} = 1$ about its major axis.

8. The region in the first quadrant bounded by $y = \dfrac{4}{x}$, $y = x$ and $y = 1$ is rotated about the y-axis. Find the volume of the solid formed.

9. A hole of radius $\frac{a}{2}$ is drilled through a solid sphere of radius a, with one edge

   of the hole passing through the centre of the sphere. The volume of the

   material removed is $\dfrac{2(3\pi-4)a^3}{n}$ where n is integer. Find the value of n.

10. Find the volume of the solid formed by rotating $y = \cos x$ around the y-axis.

    (This means the finite area above the x-axis between $x = 0$ and $x = \dfrac{\pi}{2}$).

11. Find the volume of the solid formed by rotating the area enclosed by $y = \sqrt{x}$,

    $x^2 + y^2 = 20$ and the x-axis around the x-axis.

Answers to Worksheet 4

1. $\dfrac{10\pi}{3}$ 2. 5.9698   3. $4\pi^2$ 4. $\dfrac{24\pi\sqrt{2}}{5}$

5. $\dfrac{4\pi}{15}$ 6. $\dfrac{28\pi}{15}$ 7. $\dfrac{320}{3}\pi$ 8. $\dfrac{17\pi}{3}$

9. 8   10. 3.5864   11. 28.15

Worksheet 5

1. Let R be the region in the first quadrant bounded by the graphs of $y = 2e^{-x}$

   and the line $x = k$.

   a) Find the area of R in terms of k.

   b) Find the volume of the solid formed when R is rotated $360°$ about the x-

   axis.

   c) Find the volume in part b) as $k \to \infty$.

2. Let R be the shaded region in the first quadrant enclosed by the y-axis and

   the graphs of $y = 4 - x^2$ and $y = 1 + 2\sin x$ as shown in the figure below.

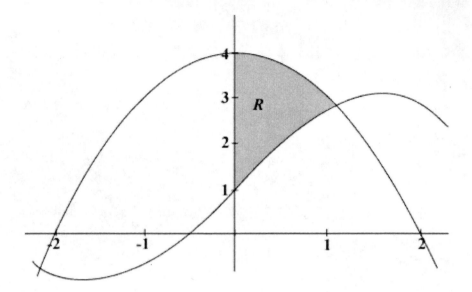

   a) Find the area of R.

   b) Find the volume of the solid generated when R is revolved about the

      x-axis.

   c) Find the volume of the solid whose base is R and whose cross sections

      perpendicular to the x-axis are squares.

3. Let R be the region in the first quadrant under the graph of $y = \dfrac{1}{\sqrt{x}}$ for

   $4 \le x \le 9$.

   a) Find the area of R.

   b) If the line $x = k$ divides the region R into two regions of equal area, what

      is the value of k?

   c) Find the volume of the solid whose base is the region R and whose

      cross-sections cut by planes perpendicular to the x-axis are squares.

4. Let R be the region enclosed by the graphs of $y = \ln(x^2 + 1)$ and $y = \cos x$.

a) Find the area of R.

b) The base of a solid is the region R. Each cross-section of the solid perpendicular to the x-axis is an equilateral triangle. Find the volume of the solid.

5. Find the volumes of the solids described:

a) The solid lies between planes perpendicular to the y-axis at $y = 0$ and $y = 2$. The cross-sections perpendicular to the y-axis are circular discs with diameters running from the y-axis to the parabola $x = \sqrt{5}y^2$.

b) The solid lies between planes perpendicular to the x-axis at $x = -1$ and $x = 1$. The cross-sections perpendicular to the x-axis between these planes are squares whose diagonals run from the semicircle $y = -\sqrt{1-x^2}$ to the semicircle $y = \sqrt{1-x^2}$.

Answers to Worksheet 5

1. a) $2\left(1 - e^{-k}\right)$ b) $2\pi\left(1 - e^{-2k}\right)$ c) 2π

2. a) 1.764 b) 30.46 c) 3.671

3. a) 2 b) $\dfrac{25}{4}$ c) 0.811 (approx.)

4. a) 1.168 b) 0.3967 (approx.)

5. a) 8π b) $\dfrac{8}{3}$

CHAPTER 17

Further Integration Techniques

There are no set methods for integration. In fact most functions are not capable of being integrated, but some techniques can be helpful.

INTEGRATION BY PARTS

Given that u and v are functions of x then

$$\int (uv)\,dx = u\int v - \int \left(\int v \cdot \frac{du}{dx} \right) dx$$

This looks complicated but a few examples will simplify matters.

1) $\int x \cos x\, dx$ Let $u = x$, $v = \cos x$

$$= x \sin x - \int \sin x \cdot 1\, dx$$

$$= x \sin x + \cos x + c$$

2) $\int x \cdot e^x\, dx$ Let $u = x$, $v = e^x$

$$= xe^x - \int e^x\, dx$$

$$= xe^x - e^x + c$$

3) $\int x\sqrt{1-x}\, dx$ Let $u = x$, $v = \sqrt{1-x}$

$$= x\left(-\frac{2}{3}(1-x)^{\frac{3}{2}} \right) - \int -\frac{2}{3}(1-x)^{\frac{3}{2}}\, dx$$

$$= -\frac{2x}{3}\left(1-x^{\frac{3}{2}} \right) - \frac{4}{15}(1-x)^{\frac{5}{2}} + c$$

At first sight it seems as though we always let u be the x term but this is not necessarily the case. In fact knowing which term to let be u and which to let be v is a matter of experience and practice.

4) $\int x \ln x \, dx$ Let $u = \ln x$, $v = x$

$$= \ln x \left(\frac{x^2}{2} \right) - \int \frac{x^2}{2} \left(\frac{1}{x} \right) dx$$

$$= \frac{x^2}{2} \ln x - \int \frac{x}{2} \, dx$$

$$= \frac{x^2}{2} \ln x - \frac{x^2}{4} + c$$

It should be noted that Integration by Parts is <u>not</u> a panacea for integration exercises since it simply replaces the original integral with another integration exercise.

Note also that it is important to judge which function should be u and which should be v because reversing the substitution will probably cause the integral to be unworkable.

A useful mnemonic is L A T E which is an abbreviation for

Logarithm, Algebraic, Trigonometric, Exponential

What this means is that, when given an integral containing a mix of the above, choose

u to be the function which comes first in this list.

e.g.   for $\int x^3 \ln x \, dx$ choose $u = \ln x$

for $\int x^2 \sin x \, dx$ choose $u = x^2$

for $\int x e^x \, dx$ choose $u = x$

for $\int e^x \sin x \, dx$ choose $u = \sin x$

This will produce the correct substitution most of the time.

An explanation of why Integration by Parts is true is that Integration by Parts is

closely linked to the Product Rule in differentiation as follows:

Let u, v, w be functions of x and let $D(\ )$ represent differentiation with respect to

x. Also let $v = D(w)$ i.e. $\int v = w$

<u>Product Rule</u>

$$D(uw) = wD(u) + uD(w)$$

Integrating both sides with respect to x we get

$$uw = \int wD(u) + \int uD(w) \quad \textbf{✳}$$

Substituting for w and $D(w)$ in ✳ produces

$u \int v = \int \left[\int vD(u) \right] + \int uv$ which when re-arranged becomes

$$\boxed{\int uv = u \int v - \int \left[\int vD(u) \right]}$$

Sometimes Integration by Parts needs to be done more than once.

For example

$$\int x^2 e^x dx = x^2 e^x - \int e^x 2x dx$$

$$= x^2 e^x - \left[2x \cdot e^x - \int e^x dx \right]$$

$$= x^2 e^x - 2xe^x + e^x + c$$

OR

To find $\int e^x \cos x dx$,

Let $\quad I = \int e^x \cos x dx$

$$= \cos x e^x - \int e^x (-\sin x) dx$$

$$I = \cos x e^x + \sin x e^x - \int e^x \cos x dx$$

$$I = e^x \cos x + e^x \sin x - I$$

$$2I = e^x \cos x + e^x \sin x \qquad \text{(By transferring } I \text{ to L.H.S.)}$$

$$I = \frac{1}{2} \left[e^x \cos x + e^x \sin x \right]$$

An example of using Integration by Parts in a volume of solid of revolution question follows on the next page.

Question: Find the volume of the solid formed by rotating the region R, bounded by $y = \cos x$, the positive x-axis and the positive y-axis, about the y-axis.

Answer:

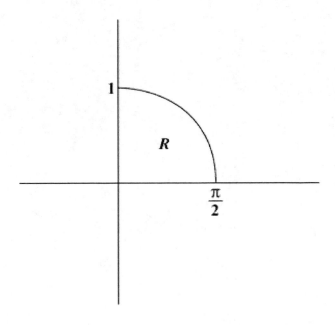

Using horizontal strips leads to the integral

$$\int_0^1 \pi x^2 dy = \pi \int (\operatorname{Arc} \cos y)^2 dy$$ which is not practical to say the least.

Using vertical strips and the shell method we obtain the result that the volume of the solid

$$= \int_0^{\frac{\pi}{2}} 2\pi xy dx = 2\pi \int_0^{\frac{\pi}{2}} x \cos x dx$$

$$= 2\pi \left[x \sin x - \int \sin x dx \right]_0^{\frac{\pi}{2}}$$

$$= 2\pi \left[x \sin x + \cos x \right]_0^{\frac{\pi}{2}} = 2\pi \left[\frac{\pi}{2} - 1 \right] = 3.586 \text{ (approx.)}$$

Sometimes Integration by Parts can be used to find the integral of a single term.

Example: $\int \ln x\, dx = \int \ln x \cdot (1)\, dx$

$$= \ln x \cdot x - \int x\left(\frac{1}{x}\right) dx$$

$$= x \ln x - \int 1\, dx$$

$$= x \ln x - x + c$$

$\int \text{Arc} \cos x\, dx$, $\int \text{Arc} \sin x\, dx$ and $\int \text{Arc} \tan x\, dx$ can all be done by using this tactic.

Substitution Methods for Integration

Sometimes it is helpful, when integrating, to rewrite the integral in terms of a different variable by means of substitution.

For example:

If $I = \int\limits_{2}^{6} x\sqrt{x-2}\,dx$ then evaluating I is not easy without the use of a calculator.

However if we let $u^2 = x - 2$ then I becomes much easier to evaluate.

(Note: Using Integration by Parts is also appropriate.)

If $\qquad u^2 = x - 2$

then $\qquad 2u\dfrac{du}{dx} = 1$

i.e. $\qquad 2u\,du = dx$

Also when $x = 2$, $u = 0$ and when $x = 6$, $u = 2$.

We can then re-write I as:

$$\int\limits_{0}^{2} \left(u^2 + 2\right)\sqrt{u^2}\,\left(2u\,du\right)$$

Note that we substitute for x, dx and the limits of integration:

$$I = \int\limits_{0}^{2} \left(2u^4 + 4u^2\right)dx$$

$$= \left[\frac{2}{5}u^5 + \frac{4}{3}u^3\right]_{0}^{2} \quad = \left(\frac{64}{5} + \frac{32}{3}\right) - 0 = 23\frac{7}{15}$$

Similarly if we wish to find $I = \int_0^4 \dfrac{\sqrt{x}}{1+x}\,dx$, then evaluating I can be simplified by

letting $u^2 = x$. Then $2u\,du = dx$ and limits of integration become $u = 0$ to $u = 2$.

$$I = \int_0^2 \frac{u}{1+u^2}\, 2u\,du$$

$$= \int_0^2 \frac{2u^2}{1+u^2}\,du$$

$$= \int_0^2 2 - \frac{2}{1+u^2}\,du$$

$$= \left[2u - 2\operatorname{Arc\,tan} u\right]_0^2$$

$$= \left(4 - 2\operatorname{Arctan} 2\right) - \left(0\right)$$

$$= 1.786 \text{ (approx.)}$$

In expressions like $\sqrt{4-x^2}$ or $9+x^2$, it is often helpful to use trigonometric

substitutions to simplify the process of evaluating an integral.

For example the identities $\cos^2\theta + \sin^2\theta \equiv 1$ and $\sec^2\theta = 1 + \tan^2\theta$ are commonly

used. In $\sqrt{4-x^2}$ if we let $x = 2\sin\theta$ note that $\sqrt{4-x^2}$ simplifies to $2\cos\theta$.

Similarly, note that if we let $x = \tan\theta$ then $9+x^2$ simplifies to $9\sec^2\theta$.

Rarely do trigonometric substitutions involve anything other than letting $x = a\sin\theta$

or $x = a\tan\theta$.

Example

Question: Evaluate $I = \displaystyle\int_0^2 \frac{1}{\sqrt{4-x^2}} dx$.

Answer: Let $x = 2\sin\theta$.

         Then $\dfrac{dx}{d\theta} = 2\cos\theta$.

         i.e. dx can be represented by $2\cos\theta\, d\theta$.

The limits of integration can be calculated by considering that

when $x = 2$, $\theta = \dfrac{\pi}{2}$ and

when $x = 0$, $\theta = 0$

         i.e. $\displaystyle\int_0^{\frac{\pi}{2}} \frac{1}{\sqrt{4 - 4\sin^2\theta}} 2\cos\theta\, d\theta$

$$= \int_0^{\frac{\pi}{2}} \frac{2\cos\theta}{\sqrt{4\cos^2\theta}} d\theta = \int_0^{\frac{\pi}{2}} d\theta = [\theta]_0^{\frac{\pi}{2}} = \frac{\pi}{2}$$

Using trigonometric substitution and by reference to the differentiation of

trigonometric functions it follows that:

$$\int \frac{1}{a^2 + x^2} dx = \frac{1}{a}\text{Arctan}\left(\frac{x}{a}\right) \text{ and}$$

$$\int \frac{1}{\sqrt{a^2 - x^2}} dx = \frac{1}{a}\text{Arcsin}\left(\frac{x}{a}\right) \text{ (assuming } |x| < a\text{)}$$

Integration by Partial Fractions

When integrating a fraction whose denominator contains terms involving x it is sometimes helpful to split up the fraction into two or three separate fractions.

For example, $\dfrac{1}{x+1}+\dfrac{2}{x-1}$ can be simplified to $\dfrac{3x+1}{x^2-1}$.

If we wish to evaluate $\displaystyle\int\dfrac{3x+1}{x^2-1}dx$ however it is clearly easier to evaluate

$\displaystyle\int\dfrac{1}{x+1}+\dfrac{2}{x-1}\,dx$.

Example

$$\int_{2}^{4}\frac{6x}{x^2+4x-5}dx$$

Note that the denominator factors into $(x+5)(x-1)$.

We therefore write $\dfrac{6x}{x^2+4x-5}$ as $\dfrac{A}{(x+5)}+\dfrac{B}{(x-1)}$ and find the values of A and B.

$$\frac{A}{(x+5)}+\frac{B}{(x-1)}=\frac{A(x-1)+B(x+5)}{(x-1)(x+5)}=\frac{6x}{x^2+4x-5}$$

Hence $A(x-1)+B(x+5)\equiv 6x$.

This is true for all x.

i.e. $\quad (A+B)x+(-A+5B)=6x+0$.

Solving, $A=5$ and $B=1$.

$$\therefore \quad \int_{2}^{4} \frac{6x}{(x+5)(x-1)}\,dx = \int_{2}^{4} \frac{5}{x+5} + \frac{1}{x-1}\,dx$$

$$= \left[5\ln|x+5| + \ln|x-1| \right]_{2}^{4}$$

$$= (5\ln 9 + \ln 3) - (5\ln 7 + \ln 1)$$

$$= 2.355 \text{ (approx.)}$$

Example

$$\int \frac{7x+1}{2x^2+x-1}\,dx = \int_{2}^{5} \frac{7x+1}{(2x-1)(x+1)}\,dx$$

$$= \int_{2}^{5} \frac{A}{2x-1} + \frac{B}{x+1}\,dx$$

$$= \int_{2}^{5} \frac{3}{2x-1} + \frac{2}{x+1}\,dx$$

$$= \left[\frac{3}{2}\ln|2x-1| + 2\ln|x+1| \right]_{2}^{5}$$

$$= \left(\frac{3}{2}\ln 9 + 2\ln 6 \right) - \left(\frac{3}{2}\ln 3 + 2\ln 3 \right)$$

$$= 3.034 \text{ (approx.)}$$

Note that

$$\frac{7x+1}{(2x-1)(x+1)} = \frac{A}{2x-1} + \frac{B}{x+1}$$

means $7x+1 = A(x+1) + B(2x-1)$

when $x = -1$, $-6 = -3B$

$$B = 2$$

and when $x = \frac{1}{2}$

$$4\frac{1}{2} = A\left(1\frac{1}{2}\right)$$

$$A = 3$$

Worksheet 1

1. Find the following integrals (no calculator).

a) $\int_1^e x\ln x\,dx$

b) $\int_{-1}^0 x\sqrt{1+x}\,dx$

c) $\int_3^5 \dfrac{8}{x^2-4}\,dx$

d) $\int_2^{2\sqrt{3}} \dfrac{4}{4+x^2}\,dx$

e) $\int \sin^3 x\,dx$ (Hint: $\sin^2 x = 1-\cos^2 x$)

f) $\int e^{\cos x}\sin x\,dx$

g) $\int_0^1 \dfrac{1}{\left(4-x^2\right)^{\frac{3}{2}}}\,dx$

h) $\int x^2 \ln x\,dx$

i) $\int x^3 e^x\,dx$

j) $\int_1^e \dfrac{2x+1}{x(x+1)}\,dx$

k) $\int e^x \sin x\,dx$

l) $\int x\left(1+x^2\right)^3\,dx$

m) $\int_0^1 2x^3\left(1+x^2\right)^5\,dx$

Answers to Worksheet 1

1. a) $\dfrac{e^2}{4}+\dfrac{1}{4}$ b) $-\dfrac{4}{15}$ c) 1.524 d) $\dfrac{\pi}{6}$ e) $-\cos x+\dfrac{1}{3}\cos^3 x$

f) $-e^{\cos x}$ g) $\dfrac{1}{4\sqrt{3}}$ h) $\dfrac{1}{3}x^3\ln x-\dfrac{x^3}{9}$ i) $x^3 e^x -3x^2 e^x +6xe^x -6e^x$

j) $1+\ln\left(\dfrac{e+1}{2}\right)$ k) $\dfrac{1}{2}\left(e^x\sin x-e^x\cos x\right)$ l) $\dfrac{1}{8}\left(1+x^2\right)^4$

m) $\dfrac{107}{14}$

Worksheet 2 (No Calculator)

1. Find the area of the region bounded by the x-axis, the line $x=2$ and the graph of $y=x\ln x$.

2. A particle moves along a horizontal line such that its velocity $v(t)=t\cos t$ where $t\geq 0$. t is measured in seconds and $v(t)$ is measured in metres per second.

   a) Find an expression for the acceleration at $a(t)$ after t seconds.

   b) Find an expression for the position $x(t)$ after t seconds given that the particle starts at a position of 0.

   c) Find the total distance travelled by the particle from $t=0$ to $t=\pi$.

3. Given a relation such that its slope $\dfrac{dy}{dx}=\dfrac{\ln x}{3y^2}$ at any point (x,y) on the graph and $y=0$ when $x=1$ find the value(s) of y when $x=e$.

4. Evaluate the following definite integrals.

   a) $\displaystyle\int_{1}^{\ln 3} xe^x\,dx$

   b) $\displaystyle\int_{0}^{1} \operatorname{Arc\sin} x\,dx$

   c) $\displaystyle\int_{0}^{\sqrt{3}} \frac{1}{\left(1+x^2\right)^{\frac{3}{2}}}\,dx$

5. The region bounded by the curve $y = x - \dfrac{1}{x}$, the x-axis, $x = 2$ and $x = 3$, is

   rotated about the x-axis. Find the volume of the figure so formed.

6. The region bounded by $y = (x-1)^2$ and $y = 1$ is rotated about the y-axis.

   Find the volume of the figure so formed.

7. Evaluate $\displaystyle\int_{e^2}^{e^3} \dfrac{1}{x \ln x - x}\, dx$.

Answers to Worksheet 2

1. $\ln 4 - \dfrac{3}{4}$

2. a) $a(t) = \cos t - t \sin t$ b) $x(t) = t \sin t + \cos t - 1$

   c) Total distance is π.

3. $y = 1$

4. a) $3(\ln 3 - 1)$ b) $\dfrac{\pi}{2} - 1$ c) $\dfrac{\sqrt{3}}{2}$

5. $\dfrac{9\pi}{2}$

6. $\dfrac{8\pi}{3}$

7. $\ln 2$

Worksheet 3 (No Calculators)

1. A particle moves along the x-axis so that its position at time t (greater than 0)

    is given by the formula $x(t) = \int_0^t u \sin u \, du$.

    a) Find the velocity and acceleration of the particle at time $t = \dfrac{\pi}{2}$.

    b) Does the particle change direction when $t = \pi$?

    c) Find the maximum position of the particle in the first four seconds.

2. Evaluate $\displaystyle\int_0^4 \dfrac{1}{1+2\sqrt{x}}\,dx$.

3. Evaluate $\displaystyle\int_{\frac{1}{3}}^{\frac{\pi+1}{3}} 2x\sin(3x-1)\,dx$.

4. Evaluate $\displaystyle\int_1^2 x(x-1)^5\,dx$.

5. Find $\displaystyle\int \sin(\ln x)\,dx$.

6. Evaluate $\displaystyle\int_0^9 \sqrt{1+\sqrt{x}}\,dx$.

7. Find $\displaystyle\int \dfrac{e^{2x}}{\sqrt{e^x - 1}}\,dx$.

8. Find the area of the region bounded by the x-axis, the line $x = 2$, and the graph of $y = x \ln x$.

9. Given a relation such that its slope $\dfrac{dy}{dx} = \dfrac{\ln x}{2y}$ at any point (x, y) on the graph and $y = 0$ when $x = 1$ find the value(s) of y when $x = e$.

10. Find the total area enclosed by $y^2 = x^2 - x^4$.

Answers to Worksheet 3

1. a) $v = \dfrac{\pi}{2}$, $a = 1$.   b) Yes   c) π

2. $\dfrac{4 - \ln 5}{2}$ 3. $\dfrac{2}{9}(\pi + 2)$ 4. $\dfrac{13}{42}$

5. $\dfrac{1}{2}\left(x \sin(\ln x) - x \cos(\ln x)\right) + c$ 6. $\dfrac{232}{15}$

7. $\dfrac{2}{3}(e^x - 1)^{\frac{3}{2}} + 2(e^x - 1)^{\frac{1}{2}} + c$ 8. $\ln 4 - \dfrac{3}{4}$ OR 0.636

9. $y = 1$ or -1 10. $\dfrac{4}{3}$

486

Improper Integrals

Improper Integrals are described to be integrals where the limits of the integral are infinite or where the function tends to infinite values. Some examples follow.

Consider $I = \int_{1}^{\infty} \frac{1}{x^2}\, dx$

$$I = \left[-\frac{1}{x} \right]_{1}^{\infty} = \lim_{t \to \infty} \left[-\frac{1}{x} \right]_{1}^{t} = \lim_{t \to \infty} \left(-\frac{1}{t} + 1 \right) = 1$$

Note that it follows that the area under $y = \frac{1}{x^2}$, above the x-axis, from $x = 1$ to an infinite value of x has a finite area, in fact 1.

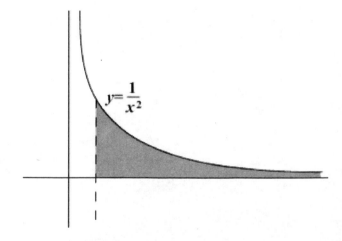

i.e. shaded region has area 1.

Similarly $\int_1^\infty \dfrac{1}{x^3}\,dx = \left[-\dfrac{1}{2x^2}\right]_1^\infty = \dfrac{1}{2}$

It soon becomes clear that $\int_1^\infty \dfrac{1}{x^n}\,dx$ where $n > 1$ has a finite value.

However $\int_1^\infty \dfrac{1}{\sqrt{x}}\,dx = \left[2x^{\frac{1}{2}}\right]_1^\infty$ which is infinite.

Similarly $\int_1^\infty \dfrac{1}{x}\,dx = \left[\ln x\right]_1^\infty$ which is also infinite.

It follows that $\int_1^\infty \dfrac{1}{x^n}\,dx$ is FINITE if $x > 1$

is INFINITE if $x \le 1$

Consider $I = \int_0^1 \dfrac{1}{x^2}\,dx$. This is also an improper integral since as $x \to 0$, $\dfrac{1}{x^2} \to \infty$.

$$I = \left[-\dfrac{1}{x}\right]_0^1 = (-1) - \left(\dfrac{1}{0}\right) \text{ which is infinite.}$$

Similarly $\int_0^1 \dfrac{1}{x^3}\,dx$ is infinite.

However $\int_0^1 \dfrac{1}{\sqrt{x}} = \left[2x^{\frac{1}{2}}\right]_0^1 = 2 - 0 = 2$ which is finite.

It follows that $\int_0^1 \dfrac{1}{x^n}\,dx$ is INFINITE if $x \ge 1$

is FINITE if $x < 1$

This means when looking at a graph of $y = \dfrac{1}{x^2}$ the vertical line $x = 1$ splits up the

area into two regions both of which have extremities tending to infinity but in which

one of the regions has finite area and the other region has infinite area.

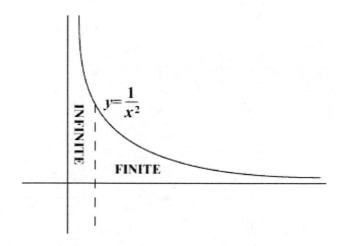

$y = \dfrac{1}{x}$ is the "dividing" function since $x = 1$ divides the area under the curve into two

regions each of which has infinite area.

Other results that defy intuition follow on the next page.

Consider the region bounded by $x = 1$ to $x = +\infty$ above the x-axis but below $y = \dfrac{1}{x}$.

We have seen already that this region has an infinite area. However consider what happens when this region R is rotated around the x-axis.

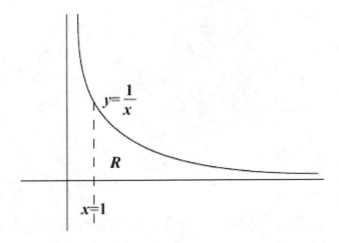

The volume of the solid of revolution so formed is

$$\int_{1}^{\infty} \pi y^2\, dx = \pi \int_{1}^{\infty} \frac{1}{x^2}\, dx$$ which we have seen already is FINITE.

In fact, volume $= \pi$.

At first sight this result seems to contradict Pappus' Theorem but Pappus' Theorem does not apply here because the centre of gravity of region R is not well-defined in co-ordinate terms.

Another result which is perhaps counter-intuitive follows on the next page.

Consider the region "bounded" by the x-axis, the y-axis, $y = \dfrac{1}{\sqrt{x}}$, and $x = 1$.

Call it R.

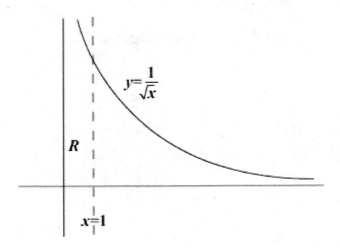

As we have seen before the area of $R = \displaystyle\int_0^1 \dfrac{1}{\sqrt{x}}\,dx$ which is finite; in fact area is 2.

Now consider what happens when region R is rotated around the x-axis. The volume of this solid of revolution is

$$\int_0^1 \pi y^2\,dx = \pi \int_0^1 \dfrac{1}{x}\,dx \text{ which is infinite.}$$

Here Pappus' Theorem is not contradicted because the centre of gravity of region R is $\left(\dfrac{1}{3}, \infty\right)$ and hence the distance travelled by the centre of gravity around the x-axis is $2\pi\infty$.

Worksheet 4

1. Evaluate $\displaystyle\int_0^1 \frac{1}{\sqrt{1-x}}\,dx$.

2. Evaluate $\displaystyle\int_1^\infty \frac{1}{1+e^x}\,dx$.

3. Calculate the volume of the solid formed by rotating the area under the curve

   $y = \dfrac{1}{1+x^2}$ for $x \geq 0$ around the x-axis.

4. Evaluate $\displaystyle\int_0^\infty xe^{-x}\,dx$

5. Evaluate $\displaystyle\int_0^\infty x^3 e^{-x^2}\,dx$.

6. a) Does $\displaystyle\int_0^{\frac{\pi}{2}} \tan x\,dx$ have a finite value?

   b) Does $\displaystyle\int_0^{\frac{\pi}{2}} \sec x\,dx$ have a finite value?

7. $f(x) = \dfrac{1}{x^2}$ is a function such that $f(a) > 0$ for all a.

   Yet $\displaystyle\int_{-1}^1 \frac{1}{x^2}\,dx = \left[-\frac{1}{x}\right]_{-1}^{+1} = \left(-\frac{1}{1}\right) - \left(-\frac{1}{-1}\right) = -2$.

   Explain the apparent contradiction.

8. a) Evaluate $\displaystyle\int_{2}^{6}\frac{x}{\sqrt{x-2}}dx$.   b) Evaluate $\displaystyle\int_{0}^{3}\frac{1}{(x-1)^{\frac{2}{3}}}dx$.

c) Evaluate $\displaystyle\int_{0}^{4}\frac{1}{\sqrt{4-x}}dx$.   d) Evaluate $\displaystyle\int_{0}^{\infty}e^{-x}\cos x\,dx$.

9. Find the smaller area cut from the ellipse $\dfrac{x^2}{4}+\dfrac{y^2}{9}=1$ by the line $3x+y=6$.

10. a) $x=k$ is a vertical line which divides into two equal parts, the area in the

first quadrant under the graph of $y=e^{-x}$ as $x\to\infty$. Find the value of k.

b) The two equal areas referred to in part a) are rotated about the x-axis.

Are the resulting volumes equal?

11. R is the region "bounded" by the x-axis, the y-axis, $y=\dfrac{1}{\sqrt{x}}$ and $x=1$.

Find the volume of the solid of revolution formed when R is rotated around

the y-axis. Does Pappus' Theorem apply?

Answers to Worksheet 4

1. 2

2. 0.31326

3. 2.4674

4. 1

5. $\dfrac{1}{2}$

6. a) No   b) No

7. There is a discontinuity when $x=0$.

8. a) $\dfrac{40}{3}$ b) 6.78 c) 4 d) 0.5

9. $6\left(\dfrac{\pi}{4}-\dfrac{1}{2}\operatorname{Arcsin}\dfrac{3}{5}-\dfrac{2}{5}\right)$

10. a) $\ln 2$ b) No $\dfrac{3\pi}{8}$ and $\dfrac{\pi}{8}$

11. $\dfrac{4\pi}{3}$. Yes, Pappus' Theorem applies.

Index

ISBN 1-41205874-0